from source to use
Energy

from source to use
Energy

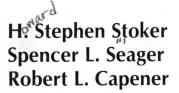
H. Stephen Stoker
Spencer L. Seager
Robert L. Capener

Weber State College

Scott, Foresman and Company • Glenview, Illinois
Dallas, Tex. • Oakland, N.J. • Palo Alto, Cal.
Tucker, Ga. • Brighton, England

Library of Congress Catalog Number: 74-78225
ISBN: 0-673-07947-3

Preface

In spite of predictions, since the early '60s, of an impending energy crisis and more current brownouts and shortages of gasoline, heating oil, and diesel fuel, many Americans seem unconvinced that the energy crisis is anything more than a temporary situation contrived by the major oil corporations and politicians. The resumption of the flow of gasoline (at a higher price) after the lifting of the Arab oil embargo has only reinforced suspicions that the crisis was contrived. The hard reality, however, is that present energy sources are limited and are being consumed at an ever increasing rate.

How serious is the situation? Exactly what resources do we have, and how soon can we develop other sources of energy? In this book we have compiled the facts from which students can derive answers to these questions. We have used a wide variety of sources and made a conscientious effort to remain objective in presenting the data. We are not attempting to place blame but rather to provide students with the data from which they can draw their own informed conclusions. Our objectivity, however, should not be interpreted as a lack of concern. We believe the energy crisis is a very serious reality that has far-reaching implications and that will not be solved quickly or simply. A solution will be the result of an extensive cooperative effort involving scientists, industrialists, economists, politicians, and especially the individual energy consumers in the United States and other nations.

Our approach in this book is the one we have found to be most successful in presenting energy problems to our students. In Part One we trace the developments that led to the energy crisis (Chapter 1); present a brief discussion of the nature of energy, including the concepts of spontaneity and efficiency, and the units used to express quantities of energy and power (Chapter 2); and examine current energy use according to use areas and energy sources (Chapter 3). Part Two consists of Chapters 4-9, which constitute the majority of the book. Chapter 4 consists of a general overview of current and proposed energy sources. Chapters 5-9 contain detailed discussions of each of these sources. Each of these chapters includes the following topics: the origin of the energy source, the processes involved in extracting useful energy from the source, reserves of the source, the future potential of the source, problems related to using energy from the source, and ways to alleviate the problems. Part Three contains only one chapter dealing with energy conservation—a topic of vital concern to all.

Before we can make intelligent decisions concerning the energy crisis, we must become informed. For this reason, we encourage you to locate and read other related materials, including those given as suggestions at the end of each chapter. The majority of these suggested readings are presented at about the same level as this book. We have also carefully indicated our sources for all data given in tabular or graphic form. We have drawn heavily on information from government publications even though some of the data in these documents have been challenged because they were collected and reported by industries that would presumably benefit from the manipulation of such information (especially the data related to fossil fuels). We have found that agreement is rare when different individuals or organizations make estimates of such things as the extent of energy reserves or the time needed to develop new energy sources. Some projections, for example, must be based upon the assumption of continued technological development or immediate action by citizens and government alike.

H.S.S., S.L.S., R.L.C.

Contents

Chapter 6 Natural Gas

Chapter 7 Coal

Chapter 8 Nuclear Energy

Chapter 9 Additional Energy Sources and Improved Energy Utilization

from source to use
Energy

Part 1

Energy:
A Servant, A Problem

Courtesy of American Airlines

Chapter 1

The Origin
of the Energy Crisis

The current per capita energy demands in the United States are the highest in the world. At the present time, each person in the United States uses, either directly or indirectly, an amount of energy approximately equal to 90 times the human caloric intake. This can be visualized as the equivalent of 90 slaves working for each United States citizen. All projections indicate that this per capita energy use in the U.S. will continue to increase in the future.

In the early 1970s many people became concerned about the ability of the United States to maintain such an energy-use pattern. This concern was touched off by short-term supply and demand imbalances, the impact of energy production and use on the environment, and projected long-term problems related to the depletion and ultimate exhaustion of fuel sources. In this chapter we will attempt to provide an overview of the origins and extent of these problems and concerns. Specific details and possible solutions are treated in more depth in Part Two of the text where each of the major energy sources (present and future) is discussed in detail.

The Term Energy Crisis

Collectively, the problems related to the production and use of energy in the United States have become known as the *energy crisis*. Americans have been exposed to this term with increasing frequency in the mass media. The term has different meanings to different people. Some think primarily of brownouts and the need for a guaranteed continuing source of electrical energy. Others become concerned about the possibility of gasoline rationing. Still others think primarily of environmental problems — air pollution, oil spills, and nuclear power plant locations and safety. These and other concerns are part of what is now called *the energy crisis*.

It is interesting to note that in some ways the term *energy crisis* is a misnomer, since the word *crisis* ordinarily implies that emergency steps can be or are being taken to ease the situation. This is not true for these energy problems; they have developed over many years and will require years to change. Furthermore, in spite of cries from some sources for an immediate Manhattan Project approach, the practical realities of the situation have not brought about a meaningful national crisis response. Such a response would require true emergency measures like energy rationing and enforced cutbacks in energy-using activities. The fact that localized problems such as brownouts, fuel shortages, and environmental contamination have arisen should concern us a great deal. For these problems are indications of the inadequacies in prevailing energy-use practices and policies, and they signify that more serious problems are very likely to develop in the near future.

As an aid to understanding the origins, complexity, and extent of the energy crisis, we will focus our discussion on four main topics. These topics are:

1. The historical development of energy use.
2. The use of renewable versus nonrenewable fuel supplies.
3. The relationships between energy use and environmental deterioration.
4. The projected energy demands for the future.

Information related to these four topics will be presented and then used to assess the magnitude of the energy crisis. It will become apparent that the energy crisis involves complex interactions of factors related to all four topics.

Historical Development of Energy Use

The use of fire was the first human utilization of energy in a form other than sunlight and the chemical energy derived from food. Fire provided protection against climatic changes and thus improved the chances for survival by increasing the habitable territorial range available for food-gathering activities. Fire also made cooking possible and previously unpalatable or indigestible substances were added to the human diet. The use of fire also increased the number and variety of materials available for human use. The development of pyrometallurgy led to the use of copper, bronze, and other metals in implements and tools previously made of stone, etc. Yet, even with the availability of fire, the greater part of the energy used to perform work was human energy derived from food.

The first big jump in energy use came about when animals were domesticated. This accomplishment tripled the amount of energy available to do work; the only cost was a little food that had to be gathered for the animals. However, the full impact of animal muscle power was not felt until much later when the wheel was invented and efficient harnesses were developed for animals.

The waterwheel came into use during the first century B.C. and, once again, the amount of energy available for use was significantly increased. It became possible to use waterwheels to exploit natural differences in elevation or level of water (potential energy). The wheels were used primarily for grinding and other mechanical tasks. The significance of the invention of the waterwheel is that it created the first source of useful energy that was not human- or animal-related.

It was not until about the twelfth century that the analogy between flowing water and flowing air was recognized and utilized in the windmill. Windmills provided energy in flat lands where the small elevation differences prevented the effective use of waterwheels. Their biggest disadvantage was the intermittent nature of their operation.

By the sixteenth century the waterwheel was by far the most important source of energy, and it provided the foundations for the industrialization of western Europe. However, by the eighteenth century Europe was in desperate need of new energy sources. Problems had developed because of increased energy needs and the fact that waterwheel outputs varied with the weather. Also, by this time most good industrial sites with sufficient water flow were already in use.

The beginning of the modern era of energy use coincided with the development of the steam engine. Although its power output was originally less than that of the waterwheel (see Figure 1—1), the steam engine completely transformed the economic, cultural, and social life of the period.

The steam engine was initially used in an auxiliary capacity to pump water to higher elevations in order to increase the energy output of waterwheels. It slowly increased in use until it finally became the dominant energy source of that time. By the middle of the nineteenth century steam energy was the major prime mover for the manufacturing industry. Much of the revolutionizing effect of the steam engine was the result of its being the first really mobile energy source.

The internal combustion engine entered the energy picture in the late 1800s and enlarged and accelerated the social changes that had begun with the use of steam energy. The internal combustion engine has been followed in turn by such devices as water, steam, and gas turbines. The discovery, development, and utilization of electricity were equally as important as the internal combustion engine in pushing the use of energy upward. The first power station to provide electricity for private consumers was established in 1882.

The power available for individual use has increased dramatically with time, as shown by the information of Figure 1—1. We see that the power output of basic machines has climbed more than five orders of magnitude (increased by a factor of 100,000 times) since the start of the industrial revolution in approximately 1750.

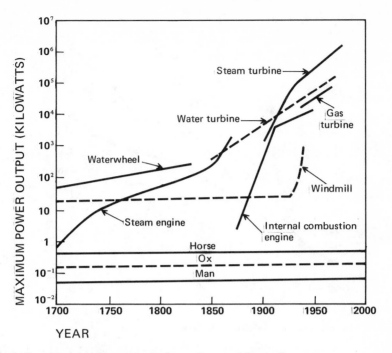

Figure 1—1.   Historical Changes in the Power Output of Energy Sources

Population Growth and Energy Use

According to the preceding brief historical discussion, the only energy used from the time of animal domestication until about 1700 was the muscle power of humans and animals plus a small amount derived from burning wood, blowing wind, and falling water. The amounts of energy used were small compared to that produced by the machines that came later (see Figure 1—1). It is significant that during this period the world's use of energy was very nearly proportional to the human population. However, after the steam engine was fully developed, energy use began to grow more rapidly than the population. Figure 1—2 provides a comparison of energy and population growth for the United States during the period 1850—1970. Note the steady increase in energy use over population increase since 1900. With the exception of an energy-use decline during the depression, the ratio of energy use to population has steadily increased. The ratio, in terms of 10^8 Btu per person, was 1 : 1 in 1860, 2 : 1 in 1920, 3 : 1 in 1970, and is projected to be 4 : 1 by 1980. This ever widening gap between energy use and population growth is one of the fundamental problems that make up the energy crisis.

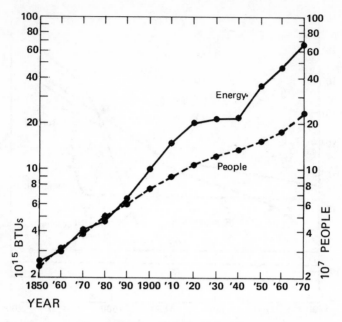

Figure 1—2.   Energy and Population Growth in the United States

Redrawn with permission from "Energy and the Environment" by John M. Fowler, published in *The Science Teacher*, 39, 9 (Dec. 1972): 12.

Renewable Versus Nonrenewable Fuels

The fuels used as energy sources have changed with time just as have the methods for obtaining energy. Into the early 1800s most useful energy was obtained from windmills, waterwheels, and the burning of wood. Wood was the fuel used in early steam engines, including those of the riverboats and railroad locomotives of the early 1800s. The use of coal did not begin until 1830. One hundred years ago wood was still the source of nearly 70% of the nation's useful energy. However, by 1900 the percent of energy obtained from wood fuel had dropped to only 20%, and coal had become the main energy source. These changes are reflected in Figure 1—3, which shows how wood as well as its replacement fuels contributed to the nation's energy requirements in the past and are projected to contribute in the future.

In this century we have gone from a wood- and coal-based energy supply to a system where three fourths of the energy comes from the combustion of petroleum products and natural gas (see Figure 1—3). Petroleum did not come on the scene until 1860, and its growth as an energy source has paralleled the increasing use of automobiles, aircraft, and other modern forms of transportation. It provided only 2% of the total energy used in 1900, 18% in 1925, and it now provides close to 40%.

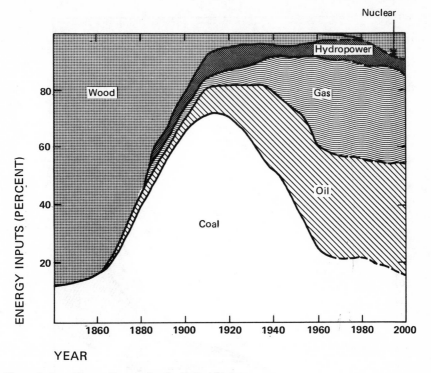

Figure 1—3. Energy Sources for the United States

Redrawn with permission from *Energy in the United States: Sources, Uses, and Policy Issues* by H. Landsberg and S. Schurr (New York: Random House, Inc., 1963), page 30.

The use of natural gas started to increase about 1930 and increased especially rapidly after World War II, when the development of large-diameter pipelines allowed the gas to be moved easily throughout the country.

Nuclear energy sources are just beginning to satisfy significant amounts of the nation's energy demands. These sources are projected to become significantly more important in the future, as shown by the upper right side of Figure 1—3.

An alternate way of depicting the change in energy sources with time is given in Figure 1—4. This figure shows the contribution of each source to the overall energy used each year and emphasizes the great increase in the total quantity of energy used. Because of the increase in total energy use, the amount of certain fuels decreases in percentage with time but increases in terms of quantity consumed. For example, more coal is used today, when the contribution to the total is approximately 20%, than in 1900, when coal supplied 70% of the nation's energy requirements.

It is of significance to the present energy problems that historical changes have taken place in the fuels used as energy sources. The shift

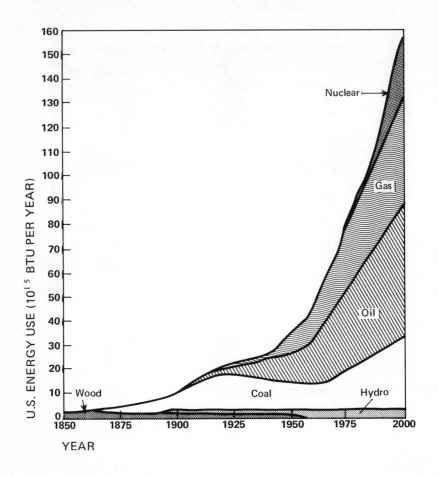

Figure 1—4.   United States Energy Sources

has been from renewable to nonrenewable fuels. While wood, wind, and water are replaceable in a relatively short time (compared to the human life-span), coal, oil, and natural gas are not. A comparison of these two types of fuels reveals the following important points:

1.   The use of renewable resources is ultimately limited by the rate of renewal rather than by the total quantity available.

2.   Theoretically, there is no limit to the rate at which nonrenewable fuels can be used but there is a finite limit on the total quantity that can be consumed.

The switch from renewable to nonrenewable fossil fuels began about 150 years ago. In this brief time-span we have become almost totally dependent on the nonrenewable variety. They are being used at a steadily increasing rate which raises concern about their ultimate depletion.

Energy and the Environment

Until the late 1960s, a fuel was generally selected on the basis of lowest overall costs; little attention was given to the possible effects of its use on the environment. Recently, however, increasing public concern about the environment has changed the concept of what is desirable. Existing environmental quality standards significantly influence the present energy supply situation. When a fuel is selected now, the environmental effects of producing, processing, and using it must be considered. This presents a complex situation since all the principal energy sources — coal, oil, gas, and nuclear — have different environmental effects, as summarized in Table 1—1.

The impact on land resources is most often related to mining activities, particularly strip or surface mining. The effects of surface mining on the environment vary with such factors as the steepness of the terrain and the amount of annual precipitation. In areas where land reclamation is not practiced, water pollution caused by acid mine drainage and silt often occurs. A 1965 survey of mining operations revealed that about 3.2 million acres of land had been damaged by surface mining. Forty-one percent of the damage resulted from coal-mining activities.

Other land problems are sometimes created as well. Underground coal mining can cause subsidence (collapse or compacting of surface land) unless the mining systems are designed to prevent deterioration and subsequent failure of abandoned mine pillars. Fuel processing often contributes large quantities of waste materials. For example, 62% of all mined coal is washed, and in the process 90 million tons of waste are produced annually. The use of coal also produces other solid wastes in the form of ash and slag. The mining and processing of uranium, the present source of nuclear energy, by either open-pit or underground methods creates similar land problems except that some of the solid wastes are radioactive.

The two water problems of greatest concern to environmentalists are water quality and water temperature. Poor water quality, either from chemical pollution or sedimentation, results in many cases from both surface and underground mining. However, the most important water problem resulting from fuel use is thermal pollution. Over 80% of all thermal pollution arises from the production of electricity. In the most efficient processes 45% of the heating value of the fossil fuel used is absorbed into cooling water. The efficiency of current nuclear power

Table 1—1.   ENVIRONMENTAL IMPACT OF ENERGY SOURCES

Impacts on Land Resource	Production	Processing	Utilization
Coal	Disturbed land	Solid wastes	Ash, slag disposal
Uranium	Disturbed land	—	Disposal of radioactive material
Oil	—	—	—
Natural gas	—	—	—
Impacts on Water Resource			
Coal	Acid mine drainage	—	Increased water temperature
Uranium	—	Disposal of radioactive material	Increased water temperature
Oil	Oil spills, transfer, brines	—	Increased water temperature
Natural gas	—	—	Increased water temperature
Impacts on Air Resource			
Coal	—	—	Sulfur oxides Nitrogen oxides Particulate matter
Uranium	—	—	—
Oil	—	—	Carbon monoxide Nitrogen oxides Hydrocarbons
Natural gas	—	—	—

Reprinted with permission from "Fuels Management in an Environmental Age" by G. A. Mills, H. R. Johnson, and H. Perry, in *Environmental Science and Technology,* 5 (Jan. 1971): 32. Copyright by American Chemical Society.

plants is worse: 55% of the produced energy enters the cooling water as heat.

Nearly 80% of all air pollution in the United States is caused by fuel combustion. About 95% of all sulfur oxides, 85% of all nitrogen oxides, and more than half of the carbon monoxide and particulates are the result of burning fossil fuels. The majority of the known low-sulfur coal is located in the western United States, far from the energy-hungry population centers of the country. This desirable low-polluting coal is thus not used because of transportation costs, and sulfur oxide pollution in urban areas is higher as a result.

Projected Energy Demands for the Future

Between 1975 and 2000 it is projected that the United States will use more energy than it has during the rest of its entire history. This can be seen from the growth curve given in Figure 1—5. The area (energy) underneath the curve for the years 1975—2000 (shaded) is

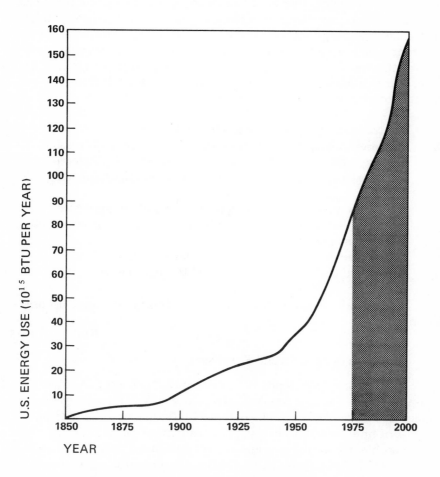

Figure 1—5.   Past and Future Energy Demands in the United States

Modified from "Energy and Power" by Chauncey Starr. Copyright Sept. 1971 by Scientific American, Inc. All rights reserved.

greater than the area under the rest of the curve. These projected increases in energy use will certainly tax our abilities to discover, extract, and refine fuels in the huge quantities necessary. The problems related to supplying such quantities are now matters of great concern to those in the energy-supplying industry. Top priority should be given to the development of new energy sources, since the exhaustion of current fuels is an eventual certainty.

Dimensions of the Energy Crisis

We have seen that the current and anticipated problems related to the energy crisis are the results of interactions between various factors. Three of these factors that have been discussed to this point are (1) an increasing per capita energy use, (2) a nearly complete dependence on nonrenewable energy sources, and (3) an increasing concern for the environment. If we add a fourth factor, economics, we can gain some idea of the dimensions of the crisis and the range of problems that we face. Some examples of the type of problems that result from interactions between these four factors are:

1. The traditional fuel sources are being depleted.
2. The reserve capacity for fuel production is rapidly decreasing.
3. New exploration activities for energy sources have decreased.
4. The construction of planned production facilities has been delayed.
5. A large number of additional production and transportation facilities need to be constructed to meet the anticipated growth in energy demands.
6. A great dependence on imports has developed.
7. Non—fossil fuel energy sources need to be developed.

Some details about each problem from this noninclusive list are given below.

Depletion of Sources and Reserves

A population that depends on nonrenewable energy (or other) resources faces a serious problem: the resources will ultimately be depleted. This depletion can be delayed by finding new deposits or sources of nonrenewable energy, but it cannot be averted. The solution to the problem is to find a renewable energy source that is large enough and renewable fast enough to satisfy the demand, or to develop essentially nonexhaustible sources such as solar or nuclear fusion energy.

As the reserves of fossil fuels diminish, their cost will increase rapidly — a matter of concern for the consumer. In addition, a short supply of fossil fuels worries the chemical industry because oil and gas feedstocks are the vital raw materials used to make petrochemicals which are important intermediates for many of today's consumer products.

Until recently, major amounts of unused production capacity existed for all major fuels. This reserve capacity in the United States has just about vanished. The problem was caused by the failure of industry to recognize throughout the 1960s that demand for energy was growing faster than the industrial capacity to produce the necessary fuels. This production capacity does not refer to available fuel resources in the ground but rather to the ability of mines, wells, and refineries to deliver these resources to market. The significance of this situation is that little flexibility exists to handle emergencies such as exceptionally long or cold winters.

Decreased Exploration

Another reason for the energy shortage is the declining rate of new exploration for fossil fuels. This decline is shown for oil and gas exploration by the data of Figure 1—6.

Fuel prices have traditionally been kept at the lowest levels possible (sometimes because of governmental regulations). These prices may have been too low to provide energy producers with the needed incentive to explore for new sources.

The natural gas industry blames its supply problems on low

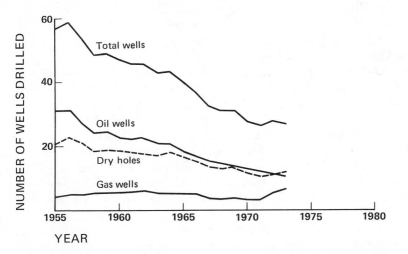

Figure 1—6.   Total U.S. Drilling Activity — Oil, Gas, Dry, and Total Wells Drilled

Data for 1972 and 1973 reprinted with permission from *Oil and Gas Journal,* April 22, 1974, table on p. 150. Data for 1955—71 reprinted with permission from *World Oil*; originally published in *Conservation and Efficient Use of Energy,* Part 3, Joint Subcommittee Hearings, House of Representatives, July 11, 1973, p. 1207, Chart 4—A.

wellhead prices which were forced on producers by the Federal Power Commission. This not only held down the generation of capital that is required for new exploration, but it eroded the incentive to invest any available funds in such activities.

The expansion of oil production has been curtailed somewhat by the activities of environmentalists. Fears of oil spills from exploratory offshore drilling and of ecological damage from the Alaskan pipeline construction have slowed efforts aimed at increasing production.

There is currently a clamor for economic incentives to encourage the finding of new reserves. Such incentives are deemed necessary by some because of increased costs due to inflation and the added expense of developing any discovered reserves in an ecologically acceptable way.

Delayed Construction and Planning of New Facilities

The construction of new production capacity to relieve the shortage described above is, with increasing frequency, behind schedule. Many of these delays are the results of concern about the possible environmental consequences of developing new production capacity. This is not to imply that such concerns are unjustified — only that they have delayed construction. Original schedules for many nuclear power plants have been greatly extended. Hearings for various licenses by the Atomic Energy Commission, state agencies, and other groups were once quite routine, but they have recently become battlegrounds for environmentalists who are concerned about radiological safety, plant locations, and thermal pollution. Schedules, which were originally planned to have plants ready in time to meet anticipated energy demands, consequently have not been met.

Enormous numbers of new facilities will be required to meet the projected energy demands of the future, whether the produced fuels are domestic or imported. Assuming the continuation of existing social, political, and economic conditions, the importing of petroleum needed to supplement domestic production will require 350 seagoing tankers, each with a deadweight (fully loaded weight) of 250,000 tons. New terminals will also have to be built, since the United States has no ports capable of receiving such supertankers. About 10 million barrels per day of new domestic refining capacity will also be required. (A barrel is equal to 42 gallons.) This will necessitate the construction of such facilities at a rate 2.5 times that of the past decade. The low-sulfur coal reserves in the western states and necessary transportation systems will have to be developed. Expanded development of underground mines in the East and Midwest will also be required. Because of inflation, the costs for these new facilities and developments will be staggering compared to those of presently existing facilities.

Dependence on Imports

Much of the slack in domestic production has been taken up with imports. The total quantities of crude oil and refined products imported annually into the United States are increasing rapidly, and natural gas imports are expected to increase dramatically in the future (see Figure 1—7).

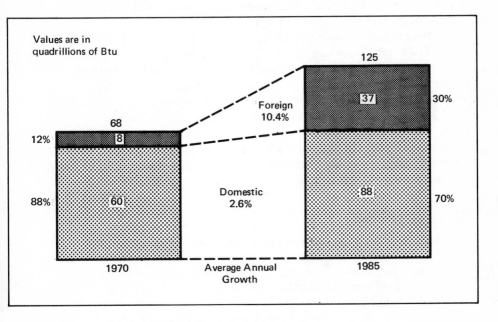

Figure 1—7.   Domestic and Foreign Fuels in the U.S. Energy Supply

Redrawn with permission from *Guide to National Petroleum Council Report on United States Energy Outlook* (Washington, D.C.: National Petroleum Council, 1972), p. 3.

Until World War II the United States was a net exporter of energy supplies. However, this is no longer true, for the United States is now not even self-sufficient. Approximately 35% of the domestic demand for crude oil and refined products is now met by imports. These imports, while helping to solve the problems of insufficient domestic production capacity, also open up a number of foreign policy questions concerning our dependence on foreign governments. The possibility that the oil-exporting nations of the Middle East might make political demands of their customers became a reality in 1973 and is still a matter for concern. Our continued dependence on large volumes of imports could also create a new financial power center in the Middle East.

Development of New Sources

The ultimate exhaustion of fossil fuel resources is certain. A very real problem we face, then, is to find replacements for them, since even technological advances in fossil fuel mining, refining, or use can only prolong the inevitable. Some of these potential sources were mentioned earlier (solar and fusion energy) and are detailed in later chapters. The development of such replacements will require extensive research and development. The amounts of time and money available for such development will be crucial and must be provided.

Many variables have contributed to the development of the energy crisis. These problems include time, the characteristics of both energy production and use, supply, demand, technology, cost, population, culture, and government. Most of these variables are not independent of each other. Demand is directly proportional to population, cultural level, and use patterns; it is indirectly proportional to cost in most cases. Supply is related to the limits of production and cost. Population and cultural level are dependent on the rate of energy supply. We may safely say that the energy crisis did not develop in a simple way, and it will not be solved in a simple way.

Suggested Readings

Cook, E., "The Flow of Energy in an Industrial Society," *Scientific American*, Sept. 1971.

Department of the Interior, "U.S. Energy — A Summary Review," Government Printing Office, Washington, D.C., Jan. 1972.

Department of the Interior, "U.S. Energy Fact Sheets — 1971," Government Printing Office, Washington, D.C., Feb. 1973.

Department of the Interior, "U.S. Energy Through the Year 2000," Government Printing Office, Washington, D.C., Dec. 1972.

Landsberg, H. H., "Low-Cost, Abundant Energy: Paradise Lost?" *Science,* April 19, 1974, pp. 247–253.

MacAvoy, P. W., P. A. Samuelson, and L. C. Thurow, "The Economics of the Energy Crisis," *Technology Review*, March/April 1974, p. 49.

Starr, C., "Energy and Power," *Scientific American*, Sept. 1971.

"U.S. Energy Outlook," National Petroleum Council, Washington, D.C., Dec. 1972.

Watt, K. E. F., "The End of the Energy Orgy," *Natural History*, Feb. 1974, pp. 16–22.

Chapter 2

The Nature of Energy

Most of us have had some experience with the energy crisis. This experience might have been in the form of an electricity shortage and resulting brownout during a hot summer day, a shortage of natural gas or fuel oil to heat homes on a bitterly cold winter day, or a shortage of gasoline to power automobiles during the vacation season. Whatever form it takes, the experience is unpleasant because it upsets our accepted way of doing things. In some cases it even creates very serious situations, such as the curtailment of farm harvests because of fuel shortages.

Because it does affect all of us, it is not surprising that the energy crisis is widely discussed by people from all walks of life. Unfortunately, these discussions often result in misunderstandings which are sometimes caused by a lack of agreement concerning the meaning of certain terms or concepts. In this chapter we present some basic information about the nature of energy which we hope will eliminate much of this disagreement and provide a common basis for the discussions contained in the chapters that follow.

Energy and Work

A definition of energy would seem to be a reasonable starting point for our discussion. *Energy* is sometimes simply defined as the capacity for doing work. However, it is more than that. In addition to work, energy may also appear in a variety of other forms, such as heat, electricity, gravitational attractions, chemical and nuclear energy, or the energy of a moving object. This broader idea is not entirely satisfactory and useful until concepts such as work, heat, etc. are understood.

Most of us have a feeling for what is meant by work. We know, for example, that it is hard *work* to push a stalled automobile, climb a steep hill, or lift heavy objects from the ground and place them in a truck. We also know that if we engage in such activities for an extended period of time certain results are inevitable: we get hungry and if we do not eat sufficient food, we lose some body weight. These results illustrate that food (a source of chemical energy) provides us with the capacity to perform the work. If we do not eat enough food, substances stored in the body are used to provide the necessary energy.

Each of the preceding examples of work fits into a commonly accepted idea: work is done when a force (f) is used to move an object some distance (d). In our examples the force was represented respectively by the push on the automobile, the push of feet against the ground, and the upward pull on the heavy objects. The distances involved were the distance the automobile was pushed (assuming a level street), a combination of the path length up the hill and the height of the hill, and the height of the truck bed above the ground (see Figure 2—1). Mathematically, the work done is equal to the product of the force and distance:

$$w = f \times d \qquad\qquad 2-1$$

The use of this equation is illustrated later in this chapter.

Some activities that are commonly called *work* do not fit into this mechanical work concept of force times distance. For example, hard studying, an activity familiar to most students, can leave a person very tired (and hungry). This results even though little work has apparently been done in the sense of a force moving an object (assuming the student sits quietly while turning the pages of a book). Of course, it must be remembered that some work is done when blood and other fluids are moved through the body and when air is inhaled and exhaled. However, careful measurements have shown that the total force-times-distance work done by the body during any activity is not equivalent to the total chemical energy that disappears. The difference shows up as energy used to carry out vital body processes, such as nerve impulse transmission and tissue repair or synthesis, and as heat which keeps the

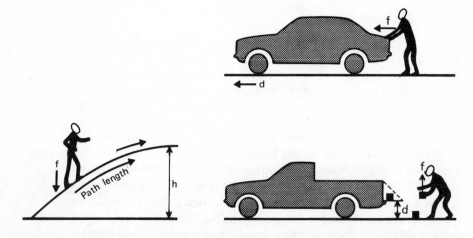

Figure 2—1. Examples of the Performance of Work

body temperature constant by replacing the heat that is lost to the surroundings.

We see from this example that energy (chemical energy, in this case) can provide the capacity to do work. But, in addition, it can cause other processes to occur which cannot be classified as mechanical work (vital body processes), and it can be transformed into other forms of energy (heat, in this case). In the hardworking student, the chemical energy did all of these things. In general, the energy of a process will be involved, with certain restrictions, in some combination of these things but not necessarily all three. One commonly encountered restriction is that a quantity of energy cannot be completely converted into work; some portion will be lost to the surroundings, usually as heat. These characteristics are common to all forms of energy and are discussed in more detail later in this chapter.

Spontaneous and Nonspontaneous Processes

Another important concept that is useful in a discussion of energy is that of spontaneous and nonspontaneous processes. A *spontaneous process* is one that occurs without external cause or stimulus, while a *nonspontaneous process* takes place only as a result of such a cause or stimulus. Some examples follow which illustrate this definition.

Suppose you are required to move a heavy, nearly spherical boulder from position A to position B. Three different situations are depicted in Figure 2—2.

In the first situation the boulder will, upon being released, roll down the hill from position A to position B. You do not have to cause it to roll down the hill; the process is spontaneous. In the second situation you must expend a little energy to push the boulder over the slight hump, but, once started, the process of rolling to position B takes place without your help. After being given a start, the process is spontaneous.

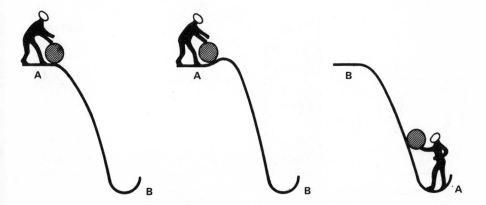

Figure 2—2.   The Problem of Moving the Boulder

In the third situation you must push on the boulder throughout the entire trip. At no time can you stop pushing and expect the boulder to continue moving from position A to position B. The process is non-spontaneous; it takes place only as a result of an external stimulus.

There are many chemical analogies to these processes. Suppose we place a small piece of white phosphorus, a kitchen match, and a glass of tap water in a rather warm room ($34°$C or $93°$F). Within a few minutes the phosphorus will burst into flames — a spontaneous process has taken place. The match, unlike the phosphorus, will not begin to burn without some help. However, if we strike the match against a rough surface and provide a little energy to start the process, the burning continues spontaneously without further outside influence. The energy provided by striking the match is analogous to the energy needed to get the boulder over the small hump in the second situation of Figure 2—2. It is often necessary to start spontaneous processes, especially those involving combustion. At the temperature of our experiment the only spontaneous process involving the water appears to be evaporation. Another process can be started by connecting wires to the opposite poles of a battery and then placing the bare ends of the wires into the water, as shown in Figure 2—3. Bubbles of gas will form on each wire as the water is decomposed into its elemental constituents, hydrogen and oxygen:

$$2H_2O \longrightarrow 2H_2 + O_2 \qquad\qquad 2\text{--}2$$

The formation of gas stops when the wires are disconnected from the battery and starts again when they are reconnected. Thus we see that the process of decomposing the water takes place only as long as energy (electricity) is supplied. The process does not start itself, and it does not continue by itself after once being started; it is nonspontaneous.

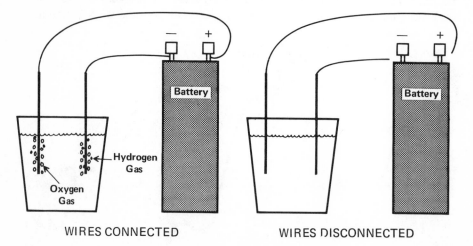

WIRES CONNECTED   WIRES DISCONNECTED

Figure 2—3.   The Electrical Decomposition of Water

The Units of Energy and Work

A number of different units are used to represent quantities of energy or work. The same units apply to both energy and work, since the two are equivalent. Table 2—1 contains some of the more commonly used units and their abbreviations.

A *Btu* is defined as the amount of heat energy required to increase the temperature of 1 pound of water by 1° Fahrenheit. The *calorie* is defined in a similar way as the amount of heat energy needed to raise the temperature of 1 gram of water by 1° Celsius. One pound is equal to about 454 grams and 1 Celsius degree is equal to 1.8 Fahrenheit degrees. The following calculation gives the number of calories equal to 1 Btu.

$$1 \text{ Btu} = (1 \text{ lb } H_2O) \ (1°F)$$

$$= (454 \text{ g } H_2O) \left(\frac{1}{1.8}°C \right)$$

$$1 \text{ Btu} = \frac{454}{1.8} \text{ cal} = 252 \text{ cal}$$

Thus, we see that 1 Btu represents more than 250 times as much energy as 1 calorie. The calorie defined above is not the "calorie" of the weight watcher, which is 1000 times larger and is frequently called a *Calorie* or a *large Calorie.*

An *erg* is a very small quantity of energy. It is very nearly the energy expended and converted into work when 0.001 grams is lifted a distance of one centimeter. Remember that one pound is equal to 454 grams and that one inch equals 2.54 cm. On this basis an erg is approximately the energy needed to lift 15 granules of table salt to a height of about ½ inch. A foot-pound is more than one million times larger. It is the energy required to lift one pound to a height of one foot. Suppose you are loading the truck mentioned earlier in this chapter. Imagine that the bed of the truck is 4 feet above the ground and that you must load 10 boxes that weigh 35 pounds each. How much work will you do while loading the truck? Since the work is equal to the force

Table 2—1.   UNITS OF ENERGY AND WORK

Unit	Abbreviation
British thermal unit	Btu
Calorie	cal
Erg	erg
Foot-pound	ft-lb
Watt-hour	whr

Table 2–2.   CONVERSION FACTORS FOR ENERGY UNITS

To Convert From	To	Multiply By	To Convert From	To	Multiply By
Btu	cal	2.52×10^2	erg	ft-lb	7.38×10^{-8}
	erg	1.05×10^{10}		whr	2.77×10^{-11}
	ft-lb	7.78×10^2			
	whr	2.93×10^{-1}	ft-lb	Btu	1.28×10^{-3}
				cal	3.24×10^{-1}
cal	Btu	3.97×10^{-3}		erg	1.36×10^7
	erg	4.19×10^7		whr	3.77×10^{-4}
	ft-lb	3.09			
	whr	1.16×10^{-3}	whr	Btu	3.41
				cal	8.59×10^2
erg	Btu	9.48×10^{-11}		erg	3.61×10^{10}
	cal	2.39×10^{-8}		ft-lb	2.66×10^3

(weight) multiplied by the distance, the work done is calculated as follows:

$$\text{Work per box} = f \times d = (35 \text{ lb})(4 \text{ ft}) = 140 \text{ ft-lb}$$
$$\text{Total work for 10 boxes} = (140 \text{ ft-lb})(10) = 1400 \text{ ft-lb}$$

or alternatively

$$\text{Total work} = \text{Total force} \times d = (350 \text{ lb})(4 \text{ ft}) = 1400 \text{ ft-lb}$$

The *watt-hour* is a unit often used to express quantities of electrical or radiant energy. We will define it in terms of its equivalence to the four previously defined units. Accordingly, one watt-hour is equal to 3.41 Btu, 859 cal, 36 billion erg, or 2660 ft-lb. These and other relationships between energy units are given in Table 2–2.

Table 2–3.   COMMONLY USED PREFIXES

Prefix	Exponential Equivalence	Numerical Equivalence
giga	10^9	1,000,000,000
mega	10^6	1,000,000
kilo	10^3	1,000
deka	10^1	10
deci	10^{-1}	1/10
centi	10^{-2}	1/100
milli	10^{-3}	1/1000
micro	10^{-6}	1/1,000,000

The results obtained earlier in the truck-loading problem can be used to illustrate the use of the conversion factors of Table 2—2. We found that 1400 ft-lb of energy were required to load the truck. How many Btu, calories, ergs, and watt-hours is this?

$$\text{number of Btu} = (\text{number of ft-lb}) \ (1.28 \times 10^{-3}) = (1400) \ (1.28) \left(\frac{1}{1000}\right)$$

$$= 1.79 \ \text{Btu}$$

$$\text{number of cal} = (\text{number of ft-lb}) \ (3.24 \times 10^{-1}) = (1400) \ (3.24) \ (1/10)$$

$$= 454 \ \text{cal}$$

$$\text{number of erg} = (\text{number of ft-lb}) \ (1.36 \times 10^{7}) = (1400) \ (1.36) \ (10,000,000)$$

$$= 19,040,000,000 \ \text{erg}$$

$$= 1.90 \times 10^{10} \ \text{erg}$$

$$\text{number of whr} = (\text{number of ft-lb}) \ (3.77 \times 10^{-4}) = (1400) \ (3.77) \left(\frac{1}{10,000}\right)$$

$$= 0.528 \ \text{whr}$$

It is often necessary to express very large or very small quantities of energy. The general practice is to do this by attaching prefixes to units such as those of Table 2—1. The more commonly used prefixes and their numerical equivalents are given in Table 2—3.

According to this table, a megawatt-hour (mwhr) equals one million (10^6) watt-hours, a kilowatt-hour (kwhr) equals one thousand (10^3) watt-hours, and a kilocalorie (kcal) is one thousand calories.

Forms of Energy

Energy is found in a variety of different forms such as that present in food, gasoline, and the falling weight of a pile driver. Some forms of energy can easily be changed into other forms. Gasoline is burned to release heat which in turn provides the energy to move an automobile. The energy expended in rubbing cold hands together produces welcome warmth.

The various forms of energy are commonly classified into two categories — kinetic and potential energy. *Kinetic energy* is that form characteristic of moving objects. The amount of kinetic energy associated with an object depends upon the mass and velocity of the object as given by equation 2—3. The energy calculated using this equation will be in ergs when the mass and velocity are in grams and centimeters per second, respectively.

$$KE = \tfrac{1}{2}mv^2 \qquad\qquad \text{2—3}$$

We see from equation 2—3 that any differences in kinetic energy for objects of equal mass are due to differences in velocity. Similarly, any differences in the kinetic energy of objects moving at the same velocity are the results of differences in mass. These relationships are illustrated in Table 2—4.

Notice that doubling the velocity of objects with the same mass causes a fourfold increase in the kinetic energy while doubling the mass of objects with the same velocity merely doubles the kinetic energy.

Potential energy is energy resulting from the relative position of interacting objects to each other. The interactions are either attractions (such as those characteristic of unlike electrical charges, unlike magnetic poles, and gravitational forces) or repulsions (between like electrical charges or magnetic poles). Thus, a book held above a desk has potential energy by virtue of its position and gravitational attraction. This energy can be changed into other forms by dropping the book onto the desk.

According to another useful definition potential energy is looked upon as being any stored energy that can be converted into kinetic energy or work. According to this definition, the following are all examples of potential energy: the chemical energy of gasoline, the electrical energy of a battery, the energy of a stretched rubber band, and the energy of water stored behind a dam.

The potential energy resulting from gravitational attractions is easily calculated using equation 2—4, where m is the mass of an object, g is the gravitational acceleration, and h is the height of the object above the earth.

$$PE = mgh \qquad\qquad \textbf{2—4}$$

The weight of an object can be substituted for the product mg and equation 2—5 results:

$$PE = wh \qquad\qquad \textbf{2—5}$$

Table 2—4.   KINETIC ENERGY OF MOVING OBJECTS

Mass (grams)	Velocity (cm/sec)	Velocity^2 (cm^2/sec^2)	Kinetic Energy (ergs)
10	5	25	250
10	10	100	1000
10	10	100	1000
20	10	100	2000

When the weight is given in pounds and the height in feet, the potential energy calculated using equation 2—5 is in foot-pounds.

Table 2—5 contains examples of the forms in which energy appears. Most of these are discussed in detail in later chapters.

Energy Production

Where does energy come from? An immediate answer to this question might include any of the sources given in Table 2—5. However, experiments and observations made more than a century ago and verified many times since indicate that energy doesn't come from anywhere; it is already here. To put it a little more scientifically, energy cannot be created nor destroyed but only changed in form. A modification to this law will be made when nuclear energy sources are discussed, but it is generally accepted as being true for nonnuclear processes.

This law of conservation of energy leads us to some interesting conclusions. For example, an electrical generator is not really an energy generator; it merely changes one form of energy into the electrical form. Much of the energy we use has been changed from one form into another before final use. In Figure 2—4 a typical series of steps is represented for the conversion of chemical potential energy into heat and light.

We notice in this process that the original energy was released by the spontaneous burning of coal. A consideration of other energy sources reveals that in every case useful energy can be derived only when a spontaneous process takes place with the evolution of energy in some form. Table 2—6 contains examples of energy sources and the spontaneous processes that release the energy. Details of these processes are discussed in later chapters.

The liberation of energy during a spontaneous process obviously leaves the products of the process with less energy than the starting materials. For example, gasoline combustion can be represented by the

---

Table 2—5.   FORMS OF ENERGY AND REPRESENTATIVE SOURCES

Kinetic	Potential
Blowing wind	Gasoline, fuel oil, food (chemical energy)
Bodies in motion (an automobile)	Battery (electrical energy)
Falling weight (an avalanche)	Sunlight (radiant energy)
Ocean tides	Uranium (nuclear energy)
Running water	Water behind a dam

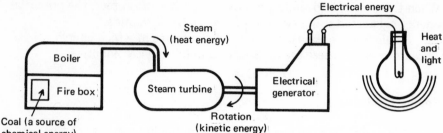

Figure 2—4.   Energy Conversion

following reaction for heptane (C_7H_{16}), one of many components found in gasoline:

$$C_7H_{16} + 11\,O_2 \longrightarrow 7CO_2 + 8H_2O + heat \qquad\qquad \textbf{2-6}$$

The law of conservation of energy requires that the total energy contained in gasoline (C_7H_{16}) and oxygen gas (O_2) must be equal to the heat liberated plus the energy contained in carbon dioxide (CO_2) and water (H_2O).

Another illustration of this idea is given in Figure 2—5 where part of the potential energy of falling water is converted into electrical energy. Some potential energy is retained by the water since it continues to flow under the influence of gravity down the river toward sea level.

Energy Utilization

The oceans of the world contain huge amounts of thermal energy. This is evidenced by the fact that ocean water can be cooled to lower temperatures and even frozen by the removal of heat. Why don't we use this vast reservoir of thermal energy to alleviate energy shortages? Heat from the ocean could be used to change water into steam and the steam could be used to drive a turbine which would do useful work. The idea

Table 2—6. ENERGY FROM SPONTANEOUS PROCESSES	
Source	Spontaneous Process
Atomic reactor	Nuclear fission
Battery	Chemical reactions
Fossil fuels (coal, oil, etc.)	Combustion
Hot water springs, geysers, etc.	Heat flows from hot to cold regions
Sunlight	Nuclear fusion
Water behind a dam	Water flows downhill

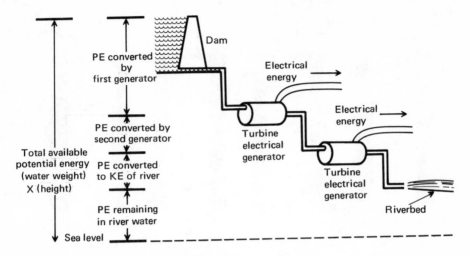

Figure 2–5.   **Electrical Energy from Falling Water**

sounds fine until we remember that heat will not spontaneously flow from the cool ocean water into the warmer water of a boiler. In fact, the ocean water would cool any warmer water in the boiler. The hottest boiler water obtainable by the proposed process would be the same temperature as the ocean water. This example illustrates the point made earlier that usable energy can be obtained only from spontaneous processes that give up energy in some form.

Suppose a spontaneous process does take place and usable energy is given off. Do other natural restrictions exist that affect the utilization of the energy? Experimental observations indicate that at least one such restriction does exist.

Consider the operation of an automobile internal combustion engine as represented in a simplified way by Figure 2–6. The heat liberated during the combustion of the fuel in the power stroke (C) causes the gaseous combustion products to expand and push on the piston. In this way the thermal energy is converted into kinetic energy which, through appropriate mechanical linkages, causes the automobile to move. The exhaust gases released at the end of the process are still quite hot and therefore contain appreciable amounts of thermal energy. How much of this remaining energy can be removed? The exhaust gases will not spontaneously lose enough heat to become cooler than their surroundings. We see then that the temperature of the medium accepting the exhaust gases places a limit on the amount of energy that can be removed. In the case of the engine depicted in Figure 2–6, this temperature is the temperature of the exhaust gases themselves, since no mechanical energy is extracted beyond step C. A little more energy could be removed and used if, for example, the exhaust gases were

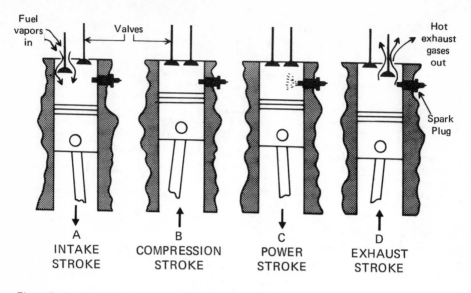

Figure 2—6.   Operation of an Internal Combustion Engine

passed through an appropriate heat exchanger and used to heat the interior of the automobile on cold days.

The maximum percentage of the available energy that can be obtained from a process and used is called the *theoretical efficiency*. In the case of heat engines, such as that of Figure 2—6, it can be calculated by using the following expression:

$$\mathscr{E}_{max} = \frac{T_1 - T_2}{T_1} \times 100 \qquad\qquad 2\text{--}7$$

In this equation the temperatures must be in absolute units which are obtained by adding 273 to the Celsius temperature. In our example of Figure 2—6, the initial temperature, T_1, is that of the gases during the power stroke. The final temperature, T_2, is that of the exhaust gases. In an internal combustion engine the burning fuel reaches a temperature of about $2760°C$ ($5000°F$) and the exhaust temperature is $815°C$ ($1500°F$). The maximum efficiency of the engine is

$$\frac{(2760 + 273) - (815 + 273)}{(2760 + 273)} \times 100 = \frac{3033 - 1088}{3033} \times 100 = \frac{1945}{3033} \times 100 = 64.1\%$$

Theoretically, this engine can make available for use almost two thirds of the energy released by the burning fuel. The remaining 35.9% passes into the surrounding atmosphere as hot exhaust. The efficiency of heat engines increases with decreasing exhaust temperatures, T_2, and becomes 100% only at absolute zero ($T_2 = 0$), an unattainable temperature.

The efficiency of energy utilization processes can be defined in a more general way that includes all processes and not just heat engines. In this definition the efficiency is equal to the percentage of available energy that is actually converted into useful work.

$$\mathscr{E}_{actual} = \frac{\text{utilized energy}}{\text{available energy}} \times 100 \qquad \textbf{2--8}$$

This definition is more useful than that given in equation 2—7 because it accounts for the unavoidable losses of energy (often due to friction) that accompany real processes.

Suppose enough gasoline is burned in an engine to liberate 100,000 ft-lb of energy. According to our results, based on equation 2—7, about 64% of this or 64,000 ft-lb are available to do work. In practice it is found that an appreciably smaller amount of energy is actually available to operate the automobile (about 30,000 ft-lb). The difference between the theoretical efficiency (equation 2—4) and the actual efficiency (equation 2—8) represents heat lost to the engine cooling system and the amount of mechanical energy converted to heat and lost to the surroundings. The conversion of mechanical energy to heat is caused by friction between the moving parts of the engine and the mechanical energy transmission systems in the automobile.

Energy losses of this type accompany all energy conversion processes. Therefore, in order to obtain as great an efficiency as possible, the number of conversion steps between the primary energy source and the ultimate use should be minimized. Gas furnaces used for heating homes have two ratings. Typical values are 125,000 Btu/hr in and 100,000 Btu/hr out. According to this rating, the furnace will provide 100,000 Btu of usable heat upon burning a quantity of fuel that releases 125,000 Btu. A calculation using equation 2—8 shows this to represent an efficiency of 80%. Imagine that instead of using the heat directly, we first convert it to steam in a boiler with 88% efficiency and use the steam to run a 50% efficient turbine electrical generator. Imagine further that the electrical resistance of the wires that carry the electricity causes a 20% transmission loss as heat. What amount of usable energy arrives at the home for heating use? Figure 2—7 answers the question, and the efficiency advantage of directly using the furnace heat becomes apparent.

If you now feel that nature is somehow going to prevent 100% energy utilization, you are correct. Studies, experiments, and experience indicate that no heat engine or any other type of energy conversion device can achieve 100% efficiency. However, in future chapters we will see that some processes are more efficient than others, and that the efficiencies of processes can often be improved if proper steps are taken.

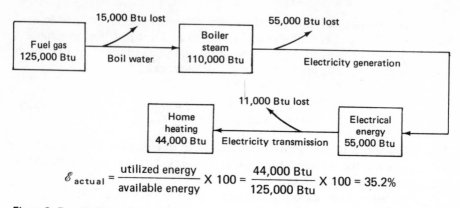

$$\mathscr{E}_{actual} = \frac{\text{utilized energy}}{\text{available energy}} \times 100 = \frac{44{,}000 \text{ Btu}}{125{,}000 \text{ Btu}} \times 100 = 35.2\%$$

Figure 2—7.   Multistep Conversion Process

Energy and Power

Suppose you cook a meal over an outdoor camp stove and find that the stove consumes about one-fourth pint of fuel in the process. Could you cook the same meal over the flame of a cigarette lighter if you had an equivalent amount of lighter fluid? The amount of available energy is the same in both cases. However, the cigarette lighter burns the fuel and liberates the energy so slowly that the heat would probably be lost to the surroundings nearly as fast as it was produced by the burning fuel. Because of these losses, it is unlikely that a pan of food could be raised to the cooking temperature by a cigarette lighter regardless of the amount of fuel available. The camp stove was capable of delivering more energy per minute to cook the food than was the cigarette lighter. The rate at which energy is used or work is done is called *power*. Thus, the camp stove developed more power than the cigarette lighter.

In general, the power developed in a process can be calculated by using equation 2—9.

$$P = \frac{\text{work done or energy expended}}{\text{time required}} \qquad\qquad 2\text{--}9$$

The work done when a 170-lb person climbs a flight of stairs to a landing 20 feet above is equal to the product of the weight lifted (170 lb) and the distance it was lifted (20 ft). Thus, 3400 ft-lb of work is done in the process. The time required to do this work does not affect the total amount of work done, but it does enter into the calculations of the power developed. If the person climbs slowly and takes 30 seconds to get to the landing, the power developed, according to equation 2—9, is

$$P = \frac{\text{work done}}{\text{time required}} = \frac{3400 \text{ ft-lb}}{0.5 \text{ min}} = 6800 \text{ ft-lb/min}$$

Table 2–7. POWER UNITS

To Convert From	To	Multiply By
British thermal units per hour	cal/hr	2.52×10^3
(Btu/hr)	erg/sec	2.93×10^6
	ft-lb/hr	7.78×10^2
	hp	3.93×10^{-4}
	w	2.93×10^{-1}
Calories per hour	Btu/hr	3.97×10^{-3}
(cal/hr)	erg/sec	1.16×10^4
	ft-lb/hr	3.09
	hp	1.56×10^{-6}
	w	1.16×10^{-3}
Ergs per second	Btu/hr	3.41×10^{-7}
(erg/sec)	cal/hr	8.58×10^{-5}
	ft-lb/hr	2.66×10^{-4}
	hp	1.34×10^{-10}
	w	1.00×10^{-7}
Foot-pounds per hour	Btu/hr	1.28×10^{-3}
(ft-lb/hr)	cal/hr	3.24×10^{-1}
	erg/sec	3.76×10^3
	hp	5.05×10^{-7}
	w	3.77×10^{-4}
Horsepower	Btu/hr	2.54×10^3
(hp)	cal/hr	6.42×10^5
	erg/sec	7.46×10^9
	ft-lb/hr	1.98×10^6
	w	7.46×10^2
Watts	Btu/hr	3.41
(w)	cal/hr	8.59×10^2
	erg/sec	1.00×10^7
	ft-lb/hr	2.66×10^3
	hp	1.34×10^{-3}

If the person climbs at a more rapid rate and gets to the top in 6 seconds, the power developed is

$$P = \frac{3400 \text{ ft-lb}}{0.1 \text{ min}} = 34,000 \text{ ft-lb/min}$$

Two common units of power are the watt and the horsepower. The watt is often used for power ratings of devices that use electrical energy. A 100-watt light bulb, for example, uses energy at the rate of 100 watt-hours per hour. The horsepower unit, used to rate mechanical devices, was defined in an interesting way. When steam engines were offered for sale to mine operators, the potential customers were interested in the number of horses each engine could replace. It was observed that, on the average, a horse could lift a total of 1320 pounds up a 1000-foot shaft in one hour. A 50% engineer's safety factor was added and the power developed by one horse was calculated as follows:

$$1 \text{ hp} = \frac{[1320 \text{ lb} + (50\%) \, (1320 \text{ lb})] \, [1000 \text{ ft}]}{(1 \text{ hr}) \left(\dfrac{60 \text{ min}}{\text{hr}} \right) \left(\dfrac{60 \text{ sec}}{\text{min}} \right)}$$

$$= \frac{(1980) \, (1000)}{3600} = \frac{550 \text{ ft-lb}}{\text{sec}}$$

The horsepower unit was then defined as a unit of power equal to 550 ft-lb/sec. Table 2—7 contains some common power units, their abbreviations, and conversion factors that relate them. These conversion factors are used the same way as those of Table 2—2 which were demonstrated earlier in this chapter.

Suggested Readings

Hirst, E., and J. C. Moyers, "Efficiency of Energy Use in the United States," *Science*, March 30, 1973, pp. 1299—1304.

Summers, C. M., "The Conversion of Energy," *Scientific American*, Sept. 1971.

Chapter 3
Energy Use Today

It is the opinion of many scientists, government officials, and energy producers that the United States (and the world) faces the prospect of a very serious energy shortage in the near future. The shortages that were first experienced during the early 1970s are thought by some to be minor compared to those on the horizon.

In order to provide an idea of the problems created by such shortages and the activities that are affected, we will devote this chapter to a discussion of the current status of energy use in the United States. Two items will be emphasized: (1) the types of uses to which energy is put today and (2) the major sources of energy.

Energy Use Per Capita

The average United States citizen today is the greatest energy user the world has ever known. Energy use is a good indicator of the standard of living. The standard of living in the United States has reached a point where the per capita use of energy is approximately 400 million Btu per year. Of course, this per capita value includes all the energy used to produce the goods and services consumed by the average citizen. It is not merely the energy used directly to run an automobile, heat a house, etc.

A comparison of annual per capita energy use among various countries is given in Figure 3—1. It is obvious that the countries with lower standards of living have lower levels of energy use. The values are for 1968 during which time the U.S. figure was 390 million Btu. During the same time an average Briton used 170 million Btu, a West German 140 million Btu, and a Brazilian 20 million Btu per year.

An annual energy use of 400 million Btu per capita represents nearly 100 times the amount of energy used by a primitive human whose requirements were primarily limited to food energy. It has been estimated that an average person must obtain about 4.4 million Btu of energy annually from food in order to survive. However, we who live in modern civilizations feel the need for much more energy than that obtainable from our food. For example, our high standard of living is

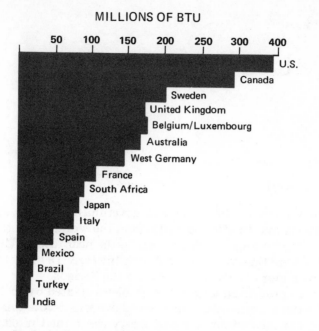

Figure 3—1.   Per Capita Energy Use, 1968

Modified from *U.S. Energy: A Summary Review* (Washington, D.C.: Dept. of the Interior, Government Printing Office, Jan. 1972), p. 1.

dependent upon the automobile; an individual who drives an automobile 5000 miles per year uses 45 million Btu of energy in the process.

Areas of Energy Use

The total annual energy use in the United States reached a level in 1974 estimated at 80.8 quadrillion (10^{15}, or a million billion) Btu. This amount is based on the 400 million Btu per capita given before and a population of slightly over 200 million. The quantity of energy represented by 80.8 quadrillion Btu per year is equivalent to the energy liberated by the daily burning of 38 million barrels of oil, 215 trillion cubic feet of natural gas, or 10.5 million tons of coal.

We will investigate this massive use of energy in terms of the amounts used in each of the following general areas: commercial, residential, transportation, and industrial. The energy used in each of these areas during 1960, 1968, and 1974 is given in Table 3—1. The 1974 figures are estimates based on observed growth rates.

The data of Table 3—1 indicate that the largest energy use occurs in the industrial area. The activities represented by the industrial area are primarily those resulting in the production of consumer goods. The

Table 3—1.   ENERGY USE BY AREA

Area of Use	Energy Used (Trillions of Btu)			Percent of National Total		
	1960	1968	1974 (est)	1960	1968	1974 (est)
Industrial	18,340	24,960	31,607	42.7	41.2	39.2
Transportation	11,014	15,184	19,288	25.5	25.2	23.9
Residential	7,968	11,616	15,370	18.6	19.2	19.1
Commercial	5,742	8,766	14,513	13.2	14.4	18.0
National Total	43,064	60,526	80,778			

Data from *Patterns of Energy Consumption in the United States,* Office of Science and Technology, Executive Office of the President, Jan. 1972, p. 6 (1974 data calculated from growth rates).

transportation area, the second largest energy user, is involved in moving people and materials from one place to another primarily through the use of automobiles, trucks, railroad trains, ships, and aircraft. The energy used in the third largest area, residential, serves primarily to provide comfort and convenience for living quarters. The fourth largest energy user, the commercial area, is made up of the activities not included in the other three areas. The details of energy use within all four areas are given later in this chapter.

The distribution of energy use among the four areas has not changed a great deal during the last fifteen years, as shown by Figure 3—2. The 1974 values are estimates based upon the compounded growth rates for each area during the 1960—68 period. These growth rates were: industrial, 3.9%; transportation, 4.1%; residential, 4.8%; and commercial, 5.4%. Two of these rates are above and two are below the national growth rate of 4.3%. Preliminary data indicate that these trends persisted into the early 1970s.

We can see that the largest energy user, the industrial area, has the lowest growth rate while the smallest user, the commercial area, has the highest growth rate. During the 1960—74 period, the share of the total annual energy use decreased in the industrial area by 3.5% (from 42.7% to 39.2%). Despite this decrease and the lowest growth rate of the four areas, the industrial use of energy still remains the dominant factor in the total energy requirements in the United States.

An investigation of the details of energy use in each of the four general areas reveals some interesting facts concerning energy use in the United States. The 1968 figures will be used in this investigation rather than the 1974 estimates because the 1968 data have been carefully analyzed. The results of this analysis were reported in January of 1972

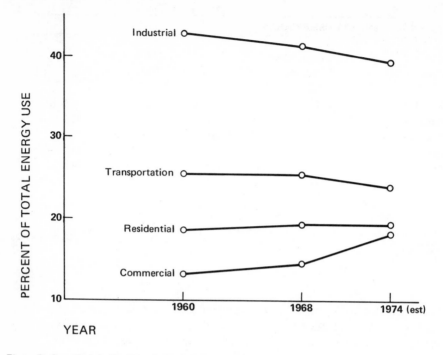

Figure 3—2.   Energy Use Distribution Among Areas

Plotted from data of Table 3—1.

by the Stanford Research Institute. Although the absolute magnitudes of the energy use figures have increased since 1968, the percentages are only slightly different at the present time.

Commercial Use of Energy

The commercial area is perhaps best defined in terms of what it is not. It is made up of all activities that are not classified as mining, manufacturing, transportation, or residential. It is therefore composed of a collection of varied, often unrelated activities and organizations such as construction companies, wholesale and retail sales outlets, finance companies, insurance companies, hotels, restaurants, schools, museums, and government institutions. The details of energy use for this area are summarized in Table 3—2.

Space and water heating are the main energy uses in the commercial area, and together they account for more than 55% of the total. However, the growth rates of these uses are lower than the area average of 5.4%. The greatest growth rates in this area are those for air conditioning with an 8.6% rate and the *other* category with a 28.0% rate. The *other* category includes commercial lighting and the energy used to drive mechanical devices such as computers, elevators, esca-

Table 3—2. ENERGY USE IN THE COMMERCIAL AREA, 1968

Type of Use	Energy Used (Trillions of Btu)	Percent of Area Total	Percent of National Total
Space heating	4,182	47.7	6.9
Air conditioning	1,113	12.7	1.8
Feedstock	984	11.2	1.6
Refrigeration	670	7.6	1.1
Water heating	653	7.5	1.1
Cooking	139	1.6	0.2
Other	1,025	11.7	1.7
Area Total	8,766		14.4

Data for energy used from *Patterns of Energy Consumption in the United States,* Office of Science and Technology, Executive Office of the President, Jan. 1972, p. 68.

lators, and office machinery. The commercial feedstock category includes asphalt and road oil which represent non-energy producing uses of petroleum, an energy source.

Residential Use of Energy

The energy in this area is used primarily to provide comfortable and convenient living conditions for the population. The details of use in this area are given in Table 3—3.

We see from Table 3—3 that space heating accounts for more than half of the energy used in this area. Energy for space heating plus the next three largest uses — water heating, refrigeration, and cooking — account for more than 83% of the total used. Air conditioning and clothes drying, the two fastest growing uses, account for only a small portion of the total use in spite of growth rates of 15.6% and 10.6%, respectively. These rates are more than double the 4.8% overall growth rate of energy use in the residential area.

Energy Use for Transportation

According to Table 3—1 the transportation area is the second largest user of energy. During 1968, the energy used for this purpose totaled 15,184 trillion Btu, a quantity representing 25.2% of the total energy used in the United States. Ninety-nine percent (15,038 trillion Btu) of this total was represented by fuel consumption. The other 1% (146 trillion Btu) was used in the manufacturing of lubricating oils and greases.

Table 3—3.   ENERGY USE IN THE RESIDENTIAL AREA, 1968

Type of Use	Energy Used (Trillions of Btu)	Percent of Area Total	Percent of National Total
Space heating	6,675	57.5	11.0
Water heating	1,736	14.9	2.9
Refrigeration	692	6.0	1.1
Cooking	637	5.5	1.1
Air conditioning	427	3.7	0.7
Lighting	412	3.5	0.7
Television operation	352	3.0	0.6
Food freezing	220	1.9	0.4
Clothes drying	208	1.8	0.3
Clothes washing	41	0.4	0.1
Dish washing	36	0.3	0.1
Other	180	1.5	0.3
Area Total	11,616		19.2

Data from *Patterns of Energy Consumption in the United States,* Office of Science and Technology, Executive Office of the President, Jan. 1972, pp. 38, 62.

Industrial Use of Energy

Energy use in the industrial area is greater than that in any of the other three general areas. More than 41% of the total energy requirements of the United States falls into this use category. Some details of energy use within the area are furnished in Table 3—4.

Table 3—4.   ENERGY USE IN THE INDUSTRIAL AREA, 1968

Type of Use	Energy Used (Trillions of Btu)	Percent of Area Total	Percent of National Total
Process steam	10,132	40.6	16.7
Direct heat	6,929	27.8	11.5
Electric motors	4,794	19.2	7.9
Feedstock	2,202	8.8	3.6
Electrolytic processes	705	2.8	1.2
Other	198	0.8	0.3
Area Total	24,960		41.2

Data from *Patterns of Energy Consumption in the United States,* Office of Science and Technology, Executive Office of the President, Jan. 1972, p. 6.

The industrial area is composed of industries engaged in manufacturing consumer products or the raw materials that go into such products. These industries are often classified according to the products or materials they produce. On the basis of such classifications, six industries account for two thirds of the energy used in the industrial area. This fact is illustrated by the data of Table 3—5.

Two of the classifications of Table 3—5, primary metals and chemicals, account for 41% of the total energy use in the industrial area. In the primary metals industry, 64% of the required energy is used in the manufacturing of iron and steel. The chemical industry is involved in the production of basic, intermediate, and end-product chemicals, including drugs and pharmaceuticals.

A little more insight into the industrial use of energy is obtained by focusing on the energy used to manufacture specific products. On this basis, sixteen industries or products are found to account for 50% of the industrial energy used. These industries and products are given in Table 3—6.

Significant End Uses of Energy

Another useful pattern of energy use can be seen by using a different method of classification. In this method energy is classified according to its end use without regard to the area in which it is used.

Table 3—5.   INDUSTRIAL AREA ENERGY USE ACCORDING TO INDUSTRIAL CLASSIFICATION, 1968

Industrial Classification	Energy Used (Trillions of Btu)	Percent of Industrial Area Total
Primary metals industry	5,298	21.2
Chemicals and allied products	4,937	19.8
Petroleum refining and related industries	2,826	11.3
Food and related products	1,328	5.3
Paper and related products	1,299	5.2
Stone, clay, glass, and concrete products	1,222	4.9
Subtotal	16,910	67.7
All other industries	8,050	32.3
Total	24,960	100.0

Data from *Patterns of Energy Consumption in the United States,* Office of Science and Technology, Executive Office of the President, Jan. 1972, p. 83.

Table 3—6.   INDUSTRIAL ENERGY USE ACCORDING TO PRODUCT

Product	Percent of Industrial Area Energy Used
Iron and steel	13.6
Petroleum refining	11.3
Paper and paperboard	5.2
Petroleum feedstock	4.9
Aluminum	2.8
Cement	2.1
Ammonia	2.0
Iron foundries	2.0
Carbon black	0.9
Grain mills	0.8
Copper	0.8
Glass	0.8
Concrete	0.7
Meat products	0.7
Soda ash	0.7
Sugar	0.7
Total	**50.0**

From *Patterns of Energy Consumption in the United States,* Office of Science and Technology, Executive Office of the President, Jan. 1972, pp. 12, 13.

When this is done, it is clear that a relatively small number of uses make significant contributions (more than 1%) to the total energy requirements of the nation. Only twelve of these uses account for 97% of the national energy requirements, as shown by Table 3—7.

Space heating is second to transportation as the largest single use of energy in the nation. Residential space heating requires more energy than commercial which, in turn, requires more than industrial. The production of process steam represents a greater use of energy than any of the other industrial activities. The category of direct heat refers to processes in which fuel is burned to provide heat which is then used directly in a process without being converted into other energy forms. It is interesting to note that energy used to drive electric motors which in turn drive machinery accounts for nearly 8% of the total national energy use. The sixth largest energy use, feedstocks and raw materials, consists of various non-energy-producing uses of fuel in forms such as lubricants and greases, asphalt and road oil, and starting materials used to manufacture other chemicals.

Table 3—7.   ENERGY REQUIREMENTS IN THE U.S. ACCORDING TO END
USE, 1968

End Use	Percent of National Energy Requirements
Transportation (fuel only, excludes lubricants and greases)	24.9
Space heating (residential and commercial)	17.9
Process steam[1] (industrial)	16.7
Direct heat[1] (industrial)	11.5
Electric motors (industrial)	7.9
Feedstocks and raw materials (commercial, industrial, transportation)	5.5
Water heating (residential and commercial)	4.0
Air conditioning (residential and commercial)	2.5
Refrigeration (residential and commercial)	2.2
Lighting (residential and commercial)	1.5
Cooking (residential and commercial)	1.3
Electrolytic processes (industrial)	1.2
Total	97.1

[1] Includes some uses for space heating; probably enough to bring the total for space heating to about 20%.

From *Patterns of Energy Consumption in the United States,* Office of Science and Technology, Executive Office of the President, Jan. 1972, p. 7.

Today's Energy Sources

Having investigated the nature of energy use in the United States, let's turn our attention to the energy sources of primary importance today. These sources are represented by both fuels and energy-producing processes. The distribution of energy production among these sources is compared for the years 1960, 1968, and 1974 in Figure 3—3. The 1974 values are estimates based on the following average 1960—68 annual growth rates: coal, 3.3%; natural gas, 5.6%; petroleum, 4.0%; hydroelectricity, 4.1%; and nuclear electricity, 47.0%. The share of the total energy requirement produced from each source changes only slightly from 1960 to 1974. The observed trend is away from the use of coal and oil, and toward the use of natural gas. The use of nuclear electricity is increasing but still makes up only a small fraction of the total.

Now that we know the main sources of energy, an important question presents itself. How is the energy from the various sources

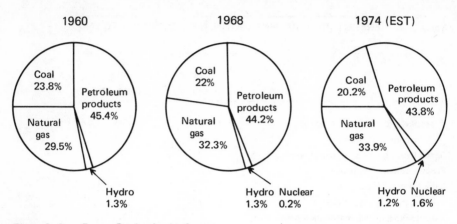

Figure 3—3.   Energy Production by Source

Data for 1960 and 1968 from *Patterns of Energy Consumption in the United States,* Office of Science and Technology, Executive Office of the President, Jan. 1972, p. 22 (1974 data calculated from 1968 data and growth rates given in text, p. 41).

distributed among the four general use areas? This question will be answered by looking at the energy distribution in two ways. (1) How much of the energy from each source is used by each use area? (2) What part of the energy requirement for each area is provided by each source? Once again, 1968 data are used because they have been completely analyzed.

Energy Distribution by Source and Use Areas

In Table 3—8 the distribution of energy from each source is given. The amount of energy involved in each use area is subdivided into direct and indirect uses. Indirect uses are those in which a fuel or process is used to generate electricity which is then used within the area.

The data of Table 3—8 show that the direct use of coal takes place almost exclusively in the industrial area; indirect coal use is significant in all areas except transportation. The industrial area is also the largest user of natural gas, accounting for more than one half of the total use. The residential area, also a significant natural gas consumer, requires more than one fourth of the supply. Petroleum is used in all areas but, as expected, its use is dominated by transportation requirements.

Calculations involving the indirect use figures of Table 3—8 reveal that 20.5% of the energy from all sources is used in an indirect way (as electricity). Furthermore, 57.3% of the electrical energy requirements are met by burning coal. The remaining demand for electricity is provided by natural gas, 26.1%; petroleum, 9.5%; hydroelectricity, 6.1%; and nuclear electricity, 1.0%.

Table 3—8.   ENERGY DISTRIBUTION AMONG USE AREAS

Energy Source and Use Area	Energy Used (Trillions of Btu)			Percent of Source Energy Used in the Area
	Direct	Indirect	Total	
Coal				
Industrial	5,616	3,215	8,831	66.3
Transportation	12	29	41	0.3
Residential	—	2,188	2,188	16.4
Commercial	568	1,698	2,266	17.0
Total	6,196	7,130	13,326	
Petroleum				
Industrial	4,474	533	5,007	18.7
Transportation	14,513	5	14,518	54.3
Residential	3,192	362	3,554	13.3
Commercial	3,389	281	3,670	13.7
Total	25,568	1,181	26,749	
Natural gas				
Industrial	9,258	1,465	10,723	54.9
Transportation	610	10	620	3.2
Residential	4,606	996	5,572	28.5
Commercial	1,845	774	2,619	13.4
Total	16,319	3,245	19,534	
Hydroelectricity				
Industrial	—	341	341	45.0
Transportation	—	4	4	0.5
Residential	—	232	232	30.6
Commercial	—	180	180	23.8
Total	—	757	757	
Nuclear electricity				
Industrial	—	58	58	44.6
Transportation	—	1	1	0.8
Residential	—	40	40	30.8
Commercial	—	31	31	23.8
Total	—	130	130	
Grand Total	48,083	12,443	60,526	

Data from *Patterns of Energy Consumption in the United States,* Office of Science and Technology, Executive Office of the President, Jan. 1972, pp. 26–30.

Table 3—9.   DISTRIBUTION OF SOURCE ENERGY

Energy Source	Percent of Area Energy Provided by Source			
	Industrial	Transportation	Residential	Commercial
Coal	35.4	0.3	18.8	25.8
Petroleum	20.1	95.6	30.6	41.9
Natural gas	43.0	4.1	48.2	29.9
Hydroelectricity	1.4	—	2.0	2.1
Nuclear electricity	0.2	—	0.3	0.4

Calculated from data given in Table 3—8.

Energy Use Distribution Among Sources

In Table 3—9 data are presented that illustrate the extent to which each energy source satisfies the energy requirements of each use area. It is apparent that natural gas, which provides nearly 50% of the needs, is the dominant energy source for the residential area while petroleum is the main source for both the commercial and transportation areas. From Table 3—8 we see that more than 66% of the coal used for energy production goes into the industrial area. However, coal provides only 35.4% of the energy requirements for the area. Natural gas is the main industrial energy source, providing 43.0% of the needs.

More details about each of these energy sources are provided later in this book, where each source is the topic of a separate chapter.

Suggested Readings

Luten, D. B., "The Economic Geography of Energy," *Scientific American*, Sept. 1971.

Office of Science and Technology, Executive Office of the President, "Patterns of Energy Consumption in the United States," Government Printing Office, Washington, D.C., Jan. 1972.

Part 2

Sources of Energy

Courtesy Tenneco, Inc., Houston, Texas

Chapter 4

Energy Sources:
General Considerations

Our increasing demand for energy and the inevitable depletion of conventional fossil fuels make it imperative that other energy sources be developed to supplement and eventually replace these fuels. It appears that such alternate sources will be more sophisticated and costly than those now in use. A number of different sources are receiving attention as possibilities for the future, but substantial scientific and engineering effort will be required to make these potential sources practical for wide-scale use. Table 4—1 contains a list of potential sources together with sources now in use. Each of these sources is dealt with in detail in later chapters.

We see from Table 4—1 that the future use of some energy sources is limited by fuel supplies (fossil fuels and fission reactors). In other cases the future use is dependent upon the geographical availability of the sources (hydroelectric, solar, tides, geothermal, etc.). Still other potential sources are in the late stages of development (fast-breeder reactors) or are yet to be proven feasible (fusion reactors).

We can use only that energy which is accessible to us — essentially, energy which is at or very near the surface of the earth. On the basis of this fact we can classify energy sources into two categories: those that provide a continuous influx of energy to the earth's surface and those that represent stored or potential energy that can be reached from the earth's surface.

Continuous Energy Influxes

The sources of the energy continuously reaching the earth's surface environment are primarily the following:

1. Solar radiation.
2. Tidal flow caused by the gravitational and centrifugal forces of the earth-moon-sun system.
3. Geothermal energy from the interior of the earth conveyed to the surface by direct conductive heating (that is, heat energy conveyed directly from molecule to molecule) and by hot water springs and volcanoes.

Table 4—1.   PRESENT AND FUTURE ENERGY SOURCES

Energy Source	Developmental Status and Prospects for Future Use
Fossil fuels	
Petroleum	Now widely used. Supplies limited — possibly exhausted in 30—40 years.
Natural gas	Now widely used. Supplies limited — possibly exhausted in 10—20 years.
Coal	Now widely used. Supplies somewhat limited — possibly exhausted in 300—500 years.
Hydroelectric	Now in use. Number of sites for future development is limited.
Solar	Now in limited use. Practicality somewhat dependent on geography, weather patterns, etc.
Nuclear	
Conventional fission reactors	Now in limited use. Low-cost fuel supply possibly exhausted in 30—40 years.
Fast-breeder reactors	Now in late stages of development. Greatly extends potential fuel supply of fission reactors.
Fusion reactors	Feasibility still to be proven. Fuel supply virtually unlimited.
Tides	Now in very limited use. Number of suitable sites for future development is limited.
Geothermal	Now in very limited use. Number of suitable sites for future development is limited.
Wind	Now in very limited use. Number of suitable sites for future development is limited.
Ocean thermal gradients	Not now used. Feasible, but dependent on geography.

A variety of methods have been used to estimate the quantities of energy delivered by each source. Data compiled by M. K. Hubbert of the U.S. Geological Survey indicate that solar radiation is by far the largest single energy source of this group, exceeding the total of the other sources by a factor of approximately 5000. Table 4—2 contains the results of this estimate.

Table 4—2.   **ENERGY INFLUXES TO THE EARTH'S SURFACE**

Energy Type	Quantity of Energy Delivered per Hour		Percent of Total Energy Delivered
	10^{12} watt-hour	10^{12} Btu	
Solar	173,000	659,130	99.980
Geothermal	32	121.9	0.018
Tidal	3	11.4	0.002
Total	173,035	659,263.3	100

Notice that one of the energy units used in Table 4—2 is the watt-hour. The number of watt-hours per hour (watt-hour/hour) is equal to and often expressed as watts because the time units cancel. Confusion sometimes results from this practice of referring to the energy per unit time in terms of the watt, a unit of power. Because of this problem and also for the sake of uniformity in this book, we will generally avoid the use of the watt-hour and use the Btu instead. This is done despite the fact that the watt-hour per hour (watt) is very commonly used to express quantities of radiant energy.

Solar Radiation

According to recent determinations made on earth and in space-craft, the average hourly energy delivered by solar radiation in space just outside the earth's atmosphere is 5315 Btu per square meter. This is the equivalent of 2.0 calories per minute per square centimeter of a surface oriented perpendicular to the radiation, or about the same as the heat generated in a typical electric kitchen oven.

The earth has an effective diameter of 1.275×10^{14} square meters exposed to solar radiation. The total amount of solar energy intercepted per hour by this area is 6.59×10^{17} Btu. This solar energy is transferred to the earth's surface environment in the form of electro-magnetic radiation which can be characterized in terms of wavelength. The shorter wavelengths correspond to higher energy radiation. The wavelength is also used to classify the radiation as visible, ultraviolet (UV), infrared, etc. The correspondence between these terms and wavelength is given in Figure 4—1.

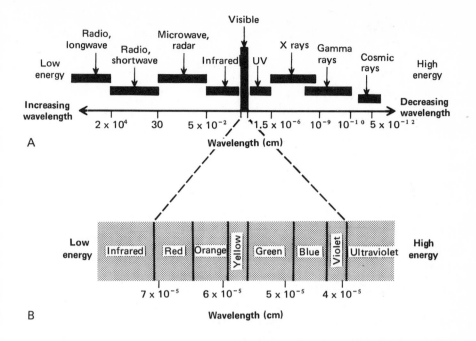

Figure 4—1. (A) The electromagnetic spectrum. There is a continuous variation in wavelength from the long-wavelength, low-energy radio waves to the short-wavelength, high-energy cosmic rays. (B) An expanded picture of the visible region of the electromagnetic spectrum. Note how small the visible section is compared with the entire spectrum.

Modified with permission from *Chemistry: Elementary Principles* by Paul F. Weller and Jerome H. Supple (Reading, Mass.: Addison-Wesley Publishing Co., Inc., 1971), p. 104.

Energy emission by a radiating body, such as the sun, takes place over a range of wavelengths. Hot objects emit shorter wavelengths than do cool ones. The effective radiating temperature of the sun is about 6000°C (11,000°F) and the energy is emitted in the form of relatively short-wavelength radiation. The relative intensity of the radiation emitted by the sun over the range of solar wavelengths is given in Figure 4—2.

Note that the maximum intensity occurs at a wavelength of 4.8×10^{-5} cm, which falls in the region of visible radiation. Also note that solar radiation covers the range from beyond ultraviolet, through visible, and into the infrared region.

As solar energy travels through the atmosphere, it interacts with atmospheric components. Some energy is absorbed by molecular oxygen (O_2), ozone (O_3), carbon dioxide (CO_2), and water vapor (H_2O). Some is reflected back into space by clouds or suspended particulate matter such as dust and water droplets. Another portion is scattered

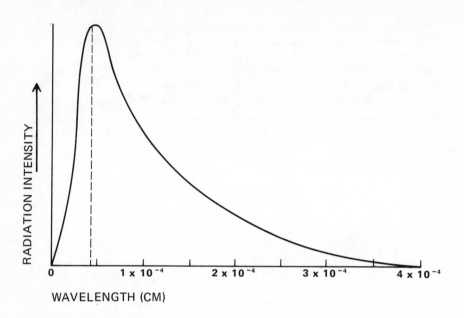

WAVELENGTH (CM)

Figure 4—2.   The Variation of Solar Radiation Intensity with Wavelength

Redrawn with permission from *General Meteorology,* 3rd ed., by H. R. Byers (New York: McGraw-Hill Book Company, 1959), p. 19.

(dispersed in all directions) by particulate matter. It is the preferential scattering of blue light (with a wavelength of about 4.3×10^{-5} to 4.8×10^{-5} cm) that gives the sky its characteristic blue color. Solar energy that is not absorbed, reflected, or scattered reaches the earth's surface where further interactions take place.

Figure 4—3 shows the percentage absorption of electromagnetic radiation by various atmospheric molecules. We see that a given molecular species absorbs only specific wavelengths and that some wavelength regions are only slightly affected by absorption.

It has been estimated that 3.08×10^{17} Btu/hr of the total incident solar radiation (6.59×10^{17} Btu/hr) is absorbed by the atmosphere. An additional 1.98×10^{17} Btu/hr is reflected and scattered back into space as short-wavelength radiation. The remainder reaches the surface of the earth, where 1.52×10^{17} Btu/hr becomes involved in the hydrologic cycle of water evaporation, precipitation, and runoff as illustrated in Figure 4—4. A small portion of 1.41×10^{15} Btu/hr is converted into the kinetic energy of winds, waves, and ocean currents and into the energy needed for the convection of water vapor. Friction eventually converts all of this energy into heat. A still smaller portion of 1.52×10^{14} Btu/hr is stored as chemical energy in plants during photosynthesis. These data can easily be compared when they are given in the form of percentages as in Figure 4—5.

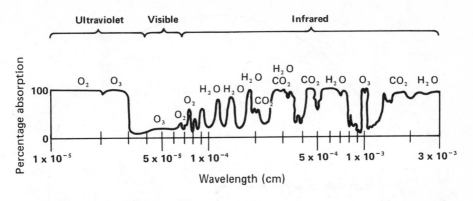

Figure 4—3. **Percentage Absorption of Electromagnetic Radiation by Atmospheric Gases**

Redrawn with permission from *Elements of Meteorology* by A. Miller and J. C. Thompson, (Columbus, Ohio: Charles Merrill Publisher, 1970), p. 69.

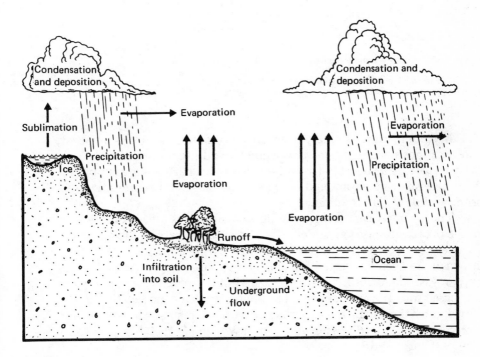

Figure 4—4. **The Hydrologic Cycle**

From *An Introduction to Environmental Sciences,* p. 174, by Joseph M. Moran, Michael D. Morgan, and James H. Wiersma. Copyright © 1973 by Little, Brown, and Company, Inc. Adapted by permission.

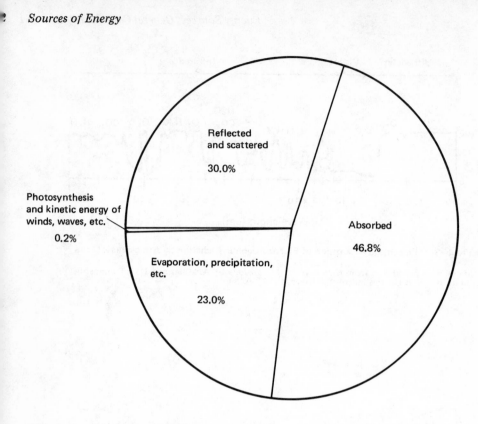

Figure 4—5.   Distribution of Incident Solar Radiation

Calculated from "The Energy Resources of the Earth" by M. King Hubbert, pp. 62–63.
Copyright Sept. 1971 by Scientific American, Inc. All rights reserved.

In spite of the seemingly insignificant amount of energy involved, photosynthesis is of extreme importance to all life on our planet. Energy stored through photosynthesis is the primary energy supply for the entire plant kingdom, and through food chains it becomes the primary supply for all animals.

The basic process involved in photosynthesis is quite complex, and all the details are not yet understood. Essentially, carbon dioxide and water are extracted from the environment by plants. These raw materials are then converted into high-energy carbohydrates and oxygen gas. The carbohydrates are stored in the plants and the oxygen is released into the atmosphere. The energy to drive the reaction is provided by the solar radiation that falls on the plants. The process is represented by the following equation in which $[C \cdot H_2O]$ is used to represent carbohydrates.

$$CO_2 + H_2O + \text{solar energy} \longrightarrow [C \cdot H_2O] + O_2 \qquad \text{4—1}$$

According to recent estimates the amount of carbon annually converted to carbohydrates by photosynthesis is 20×10^9 metric tons for land plants and 13 X 10^9 metric tons for ocean plants. One metric ton is equal to 2240 lbs.

Carbohydrates formed through photosynthesis are converted back into CO_2 and water by reactions that are essentially the reverse of photosynthesis. This occurs, for example, when plants die and decompose. The energy which was extracted from the solar radiation falling on the plants and stored in the plants is ultimately released to the environment as heat when they decay. The reaction can be written as follows:

$$O_2 + [C \cdot H_2O] \longrightarrow CO_2 + H_2O + \text{energy} \qquad \textbf{4--2}$$

Notice that oxygen is consumed in the process. When undisturbed ecological conditions prevail, photosynthesis and its reverse reactions proceed at about equal rates and a balance is achieved.

In the distant past, some plant (and animal) materials were deposited in oxygen-depleted environments, such as swamps or deep basins of stagnant water. Under such low-oxygen conditions the reactions represented by Equation 4--2 could not take place, so incomplete decomposition occurred. The products of these processes were buried under heavy layers of sand, mud, etc. and became the source of our fossil fuels — peat, lignite, coal, oil, and natural gas.

Tidal Energy

The moving water of tides is another source of continuous energy influx to the earth's surface. The tidal movement of water occurs twice daily in most parts of the world — the water rises twice and recedes twice. This movement is directly related to two factors: (1) the gravitational forces exerted on the earth by both the moon and the sun, and (2) the centrifugal force resulting from the rotational interactions of the earth, moon, and sun.

Energy can be extracted from the tidal flow of water by controlling the filling and emptying of bays or estuaries that can be closed by dams. The process is closely related to that used to obtain hydroelectricity from falling water, except that the flow is oscillatory with tides rather than unidirectional. The topic of tidal energy is covered in more detail in Chapter 9.

Geothermal Energy

The energy influx to the earth's surface environment from inside the earth takes place by two different processes — heat conduction and convection. *Conduction* is the movement of heat from molecule to molecule through a substance without any visible movement of the substance. *Convection*, on the other hand, is the movement of heat by the motion of the matter (a liquid or gas) containing the heat. These processes are described in more detail in Chapters 9 and 10. Collectively, energy that reaches the earth's surface by these two processes is called *geothermal energy*.

It has been found that the temperature of wells and mines increases with depth. At the bottom of wells drilled 2 or 3 miles deep, temperatures frequently exceed the normal boiling point of water. The rate of temperature increase with depth varies quite widely but averages about 1°C per 30 meters of depth. The average hourly heat flow to the earth's surface by conduction is 2.40×10^{-5} Btu/cm^2. The surface area of the earth is 5.10×10^{18} cm^2, and the total hourly heat flow by conduction is 1.22×10^{14} Btu.

Heat transported to the surface by convection amounts to only about 1% of that by conduction, or about 1.22×10^{12} Btu/hr. Convection processes take place primarily in volcanoes and hot-water springs.

Energy Influx Versus Outflow

During the four centuries that have passed since the invention of the thermometer, no significant changes have been detected in the overall average temperature of the earth's surface and atmosphere. This fact implies that a balance must exist between the total energy that enters the surface environment and the energy stored or lost. As we saw earlier, only a tiny fraction of the total energy reaching the earth's surface is stored through photosynthesis. The remainder is converted directly or indirectly into thermal energy at the temperature of the earth's surface. This heat is lost into space in the form of long-wavelength (infrared) electromagnetic radiation. The overall balanced system is represented by Figure 4—6.

Stored Energy

In addition to the continuous influx of energy reaching the earth's surface environment, there is a second type of energy resource available for use — stored potential energy. Nearly all of the currently used energy resources are of this type. The two major categories of stored energy are chemical and nuclear potential energy.

Chemical Potential Energy

The atom represents the limit of chemical subdivision and is the fundamental building block for the materials of the universe. At the present time, 105 different types of atoms are known. Under appropriate conditions these atoms combine in a variety of ways to produce molecules of various substances. The forces which hold together the atoms of a molecule are called *chemical bonds* and are the result of

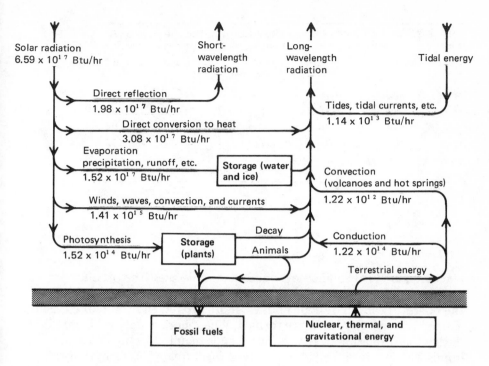

Figure 4—6.　Energy Flow Sheet for the Earth

Redrawn from "The Energy Resources of the Earth" by M. King Hubbert, pp. 62–63.
Copyright Sept. 1971 by Scientific American, Inc. All rights reserved.

interactions between the various charged parts (protons and electrons) of the combined atoms.

An input of energy is needed to break chemical bonds and separate the combined atoms. Bonds between different types of atoms require different amounts of energy. For example, the energy required to break the bond between the two hydrogen atoms of a hydrogen molecule (H—H) is nearly three times larger than that necessary to do the same thing to an iodine molecule (I—I). However, the energy needed to break a particular type of bond (such as that of the hydrogen molecule) is always the same. This necessary energy is defined as the *bond energy*. The principle of energy conservation requires that the energy input needed to break a bond must be equal to the energy given up when the bond was formed.

Many bond energies have been measured, and the values are found to depend upon the types of atoms bonded and the number of electrons involved. Some selected bond energies are given in Table 4—3, where chemical bonds are shown as dashes and each dash indicates the involvement of two electrons.

In most chemical reactions the bonds of some molecules (the

Table 4–3.   APPROXIMATE BOND ENERGIES (KILOCALORIES/MOLE)

Bond	Energy	Bond	Energy	Bond	Energy
H—H	104	N=N	100	H—O	111
C—C	83	N≡N	225	H—C	99
C=C	146	O—O	33	C—O	84
C≡C	200	O—O in O_2	118	C=O	173
N—N	40	S—S	51	C=O in CO_2	192

reactants) are broken and new bonds and molecules (the products) are formed. For example, in the photosynthetic process described earlier, the bonds of carbon dioxide and water are broken and those of oxygen gas and the carbohydrates are formed. When more energy is liberated during the formation of new bonds than is required to break the old bonds, the process is called *exothermic* (heat out). This term is quite appropriate since the net result of the reaction is the liberation of energy — usually in the form of heat. The opposite situation, in which the formation of the new bonds liberates less energy than is required to break the old bonds, gives rise to an *endothermic* (heat in) process. Thus, we can see that endothermic reactions represent energy-storing processes while exothermic reactions release the stored energy.

Most of the energy used today comes from fossil fuels (see Figure 3–3). These fuels include coal, petroleum, and natural gas, and each is a form of chemical potential energy. The original source of the energy stored in the form of fossil fuels was the sun. The fuels were produced through photosynthesis and partial decomposition, as we previously pointed out. A part of this stored solar energy is released when the fuels are burned. The burning reactions, more properly called combustion reactions, are all exothermic. In the combustion process the fossil fuel components react with oxygen to give products with lower bond energies. The net effect is the release of energy. Table 4–4 contains information about some typical reactions involved in fossil fuel combustion.

Other exothermic chemical reactions are being investigated as potential energy sources for the future. One proposal involves the use of hydrogen gas as the fuel. It is allowed to react with oxygen and liberate energy. The process releases 59.0 kcal per mole of hydrogen combusted or a rather large 5.2×10^4 Btu per pound. The reaction is:

$$2H_2 + O_2 \longrightarrow 2H_2O \qquad \qquad 4\text{--}3$$

Table 4—4. FOSSIL FUEL COMBUSTION

Fuel	Reaction Equation	Bonds Broken	Bonds Formed	Heat Evolved
Coal	$C + O_2 \rightarrow CO_2$	O—O	C=O	94 kcal/mole; 1.4×10^4 Btu/lb fuel
Natural gas*	$CH_4 + 2 O_2 \rightarrow$ $CO_2 + 2H_2O$	C—H O—O	C=O, H—O	211 kcal/mole; 2.4×10^4 Btu/lb fuel
Petroleum**	$2C_8H_{18} + 25 O_2 \rightarrow$ $16CO_2 + 18H_2O$	C—C, C—H, O—O	C=O, H—O	1303 kcal/mole; 2.1×10^4 Btu/lb fuel

*CH_4, methane, is the principal constituent in natural gas (up to 97%).

**C_8H_{18}, octane, is only one of many hydrocarbons present in petroleum.

Reprinted by permission of the publisher, from Curtis B. Anderson, Peter C. Ford, and John H. Kennedy: *Chemistry: Principles and Applications* (Lexington, Mass.: D. C. Heath and Company, 1973), p. 140.

Nuclear Potential Energy

In ordinary chemical reactions, bonds of reactant molecules are broken and those of product molecules are formed. These processes involve only the electrons of the atoms; the individual atomic nuclei remain undisturbed. These nuclei consist of particles (protons and neutrons) which are held together by nuclear forces or bonds. Under certain conditions some nuclei will undergo changes in which energy is released. These changes may occur naturally, as in the case of radioactive substances, or they may be induced, as in the hydrogen bomb. The energy released during nuclear changes is generally thousands of times greater than that released by chemical reactions involving similar weights of reactants. One process, controlled nuclear fission, is now in use as a source of energy. This and other processes are under intensive study, since some of them represent almost inexhaustible energy sources. These nuclear processes are the object of a detailed discussion in Chapter 8.

Suggested Readings

Cheney, E. S., "U.S. Energy Resources: Limits and Future Outlook," *American Scientist,* Jan./Feb. 1974, pp. 14–22.

Dyson, F. J., "Energy in the Universe," *Scientific American,* Sept. 1971.

Gates, D. M., "The Flow of Energy in the Biosphere," *Scientific American,* Sept. 1971.

Hubbert, M. K., "The Energy Resources of the Earth," *Scientific American,* Sept. 1971.

Roberts, R., "Energy Sources and Conversion Techniques," *American Scientist,* Jan./Feb. 1973, pp. 66–75.

Weinberg, A. M., "Some Views of the Energy Crisis," *American Scientist,* Jan./Feb. 1973, pp. 59–60.

Chapter 5
Petroleum

Petroleum was first used in the United States as a fuel to provide illumination. During the 1850s, illuminating oils were obtained from crude petroleum which was skimmed off ponds and streams where it had accumulated from natural seepage. The amount of petroleum available from such sources was limited and did not satisfy the demand. Petroleum was needed in large quantities. The techniques for obtaining large quantities were first demonstrated by Edwin L. Drake. On August 27, 1859, at Titusville, Pennsylvania, he completed the drilling of the first successful oil well. The immediate results of his achievement were quite dramatic: an oil boom which took on many aspects of the earlier California gold rush. In 1859, the total crude oil production amounted to only 2000 barrels; in 1860 it had risen to 500,000 barrels. (As mentioned before, a barrel is equal to 42 gallons.)

The first successful well was followed quickly by other developments in the newly born oil industry. The first oil refinery went into operation in 1861 and concentrated on producing a good odorless and smokeless kerosene — a product that was in great demand as an illuminant. In 1865 a railroad car was developed specifically to transport crude oil. The first major pipeline was completed in 1879; it extended over a distance of 110 miles.

During the early years of the twentieth century an event occurred which gave the petroleum industry its greatest impetus: the internal combustion engine was developed. Henry Ford launched his first Model T in 1908, and the demand for petroleum and petroleum products began to increase. Today, nearly one half of all energy demands in the United States are met by the use of petroleum products. Refer back to Figure 1—3 for a graphical representation of the changing role played by petroleum in the United States energy picture and to Figure 3—3 for an estimate of the current situation.

Origin and Accumulation of Petroleum

The processes by which liquid petroleum is produced in nature are not completely understood. For many years petroleum chemists have debated the question of the primary source of petroleum. Did it originally come from animals, plants, or nonbiological minerals? It is now generally believed that petroleum is derived from both plants and (to a much smaller extent) animals, but the inclusion of nonbiological components is also accepted as a possibility.

The predominantly plant origin of petroleum was proposed in the 1930s after the discovery that petroleum contains substances known as porphyrins which are clearly traceable to plant chlorophyll. In 1973, the discovery in petroleum of steroid carboxylic acids gave added acceptance to the idea of an animal contribution, since such compounds are structurally related to the bile acids found in animals.

Today, geologists generally agree that the various layers of sand and porous rock now containing petroleum deposits were at one time submerged under great bodies of water. This idea is accepted even though many such sites are now buried beneath dry land. It is postulated that countless tiny aquatic plants and animals lived and died in these bodies of water. After the plants and animals died, their remains settled to the bottom, became mixed with sand and mud, and formed layers of biological debris. As time passed the older layers were buried deeper and deeper by an increasing overburden of silt and sediment, and the pressure upon them increased significantly. This pressure, together with the heat of the earth and chemical and bacterial action, is thought to have converted the biological debris into petroleum. The resulting petroleum is generally found in porous rocks in the sedimentary layers of the earth's crust. It is usually found mixed with natural gas and sometimes salt water.

Petroleum Reservoirs

Petroleum is of great value as an energy source because significant amounts have accumulated in localized areas. Accumulation is essential since only under such conditions does extraction from the earth become economically profitable. Three geological features are essential in a suitable accumulation site for petroleum. There must be an adequate reservoir rock covered by a relatively impervious cap rock and a structural trap to confine the petroleum to a given area.

An adequate reservoir rock must be porous and permeable. The porosity of a rock is given by the percentage of void or open space it contains. Crystalline rocks may have porosities of less than 1% while some sandstones and limestones have values of 40% or higher. Permeability is a measurement of the ability of a rock to act as a conduit for moving liquids or gases. If a rock is to be highly permeable, its pores or open spaces must be interconnected. Gases and liquids migrate readily through porous, permeable rocks, and for this reason, most common reservoir rocks are sandstones or limestones.

In a suitable accumulation site the reservoir rock must be capped by rock which will prevent the upward migration of the petroleum. Cap rock is usually composed of relatively impervious shale, although some other types of rock may function equally well, provided they are impervious.

Under the conditions described above, the porous rock acts as a reservoir for gases and liquids, since their escape upward is prevented by the impervious cap rock. Natural gas is the least dense component present and tends to accumulate in the top of the reservoir. Petroleum is less dense than water and will tend to separate from any water mixed with it during the formation processes. The petroleum thus accumulates above the water and there is a separation of components into layers within the reservoir rock.

Structural Traps

The conditions for petroleum accumulation are completed when a structural trap occurs which curtails the lateral movement of petroleum and natural gas. The most common types of traps are the anticline, the fault, and the stratigraphic traps. These common types are illustrated in Figure 5–1.

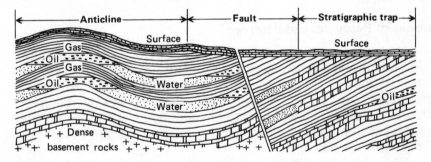

Figure 5–1.   Common Types of Structural Traps

Modified from *Plain Facts About Oil*, Dept. of the Interior, Office of Oil and Gas (Washington, D.C.: Government Printing Office, 1963), p. 2.

An anticline trap is an upward fold in the earth's strata which forms an arch. Petroleum and gas are trapped in porous structures at the tops of the resulting domes. Impervious layers are located above and below the porous layer. In most anticline reservoirs the petroleum layer is above a salt water layer which prevents any downward dispersion of the petroleum.

A fault trap is a fracture in the earth's crust which causes a porous layer to be cut off by a nonporous layer. When a fault occurs, a natural petroleum trap is formed. Petroleum may be found on either or both sides of a fault at various depths.

A stratigraphic trap results when porous layers are pinched between nonporous layers. In this type of trap there are no pronounced up-and-down folds of rock strata but only a change in porosity.

Notice that for clarity the petroleum deposits in Figure 5—1 are represented as liquid pools. A belief in this form of petroleum deposit was very popular in the early days of the petroleum industry and persists among some people today. In reality, the liquid petroleum is dispersed throughout a porous rock formation and is not collected into underground lakes or pools.

Crude petroleum extracted and brought to the surface is usually a thick, viscous liquid, and it seems unrealistic to propose that it could flow through porous rock. However, when petroleum is underground, it is subjected to higher pressures and temperatures than it is on the surface. In addition, it contains significant amounts of dissolved natural gas. All of these factors tend to make the petroleum less viscous and allow it to flow much more readily than it does on the surface.

The natural gas usually found associated with petroleum is not insignificant nor unwanted. In addition to being a valuable energy source (discussed in Chapter 6), it plays an important role in the extraction of crude petroleum. This point is discussed later in this chapter.

Petroleum Resources and Reserves

Modern technology makes it possible to obtain petroleum from both onshore and offshore locations. For this reason, realistic estimates of United States petroleum resources must include those located offshore. It is estimated that the total area included in United States offshore lands out to a water depth of about 600 feet (the continental shelf) is equal to 850,000 square miles — an area nearly one fourth that of the onshore land.

The total United States petroleum resource base, which is the petroleum originally contained in both onshore and offshore areas, is estimated to be 2830 billion barrels. Of this amount, 1360 billion barrels or 48.1% is located offshore.

The total cumulative amount of petroleum removed from the resource base up to January 1, 1971, is estimated to be 93 billion barrels. The annual production rate since then has averaged about 3.2 billion barrels. Therefore, the total amount of petroleum removed from the resource base through 1974 is about 106 billion barrels, and the original amount of petroleum present has only been decreased by about 4%.

Despite this small withdrawal from the large resource base, we continually hear that the United States petroleum reserves are quite small. This apparent inconsistency is caused by the meaning of the term *reserve*. Reserves comprise that portion of the resource base which geological or engineering data demonstrate can be recovered with reasonable certainty under existing economic and operating conditions. The resource base, on the other hand, represents the total amount of petroleum physically present regardless of whether or not recovery is feasible. The size of United States petroleum reserves is indicated in Figure 5—2, which includes an analysis of the petroleum resource base in terms of the four subdivisions: cumulative production, proven reserves, probable reserves yet to be proven, and resources not recoverable under present economic and technological conditions. Note the

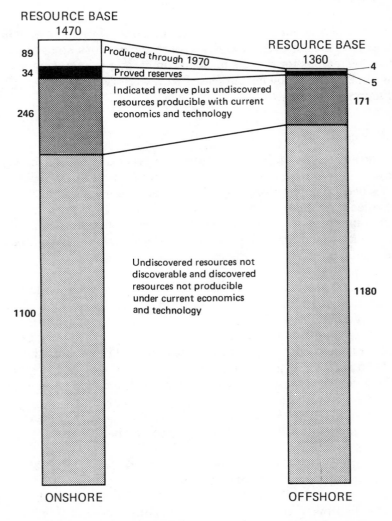

Figure 5—2.   United States Crude Petroleum Resources (billions of barrels)

Redrawn from *U.S. Energy: A Summary Review*, Dept. of the Interior (Washington, D.C.: Government Printing Office, January 1972), p. 27.

small portion of the resource base that is classified as proven reserves — an amount less than the cumulative production.

An analysis of the data of Figure 5—2 reveals several significant facts:

1. The majority of the petroleum recovered through 1970 came from onshore locations — 95.7% onshore, 4.3% offshore.
2. The majority of the proven reserves are located onshore — 87.2% onshore, 12.8% offshore.
3. A significant amount of the potential reserves are located offshore — 41.0% offshore, 59.0% onshore.

We see that petroleum production and exploration for new petroleum deposits have been heavily oriented to onshore locations up to the 1970s. The reasons for this are made apparent later in this chapter when some of the problems associated with increased offshore production are discussed.

Proven Reserves

The magnitude of proven petroleum reserves fluctuates from year to year. Reserves are increased when explorations result in probable reserves becoming proven reserves. Decreases in reserves are the results of petroleum production during the year. In Figure 5—3 the yearly variation in proven reserves since 1940 is shown as well as the magnitude of the two factors that contribute to the variation. Several trends are indicated by the data of Figure 5—3:

1. The total proven reserves steadily increased following World War II until about 1960. They then remained essentially constant until the middle sixties when they began to decrease — except for the Alaskan North Slope addition.
2. Petroleum production steadily increased until it maximized in 1970 and then began to decrease.
3. New reserve additions were highest during the period around 1950 and have slowly decreased since then.
4. During the 1940—60 period new additions were usually greater than production. The opposite has generally been true since 1960.

An estimate of the significance of proven reserves can be obtained by determining the ratio of total proven reserves at year-end to production during the year. This ratio, known as the *life index*, gives the estimated lifetime of the proven reserves if we assume that no additional reserves are found and no production increases take place. This ratio, as given in Figure 5—4, was calculated from the data of Figure 5—3.

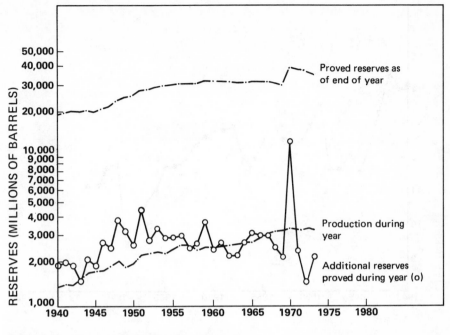

Figure 5—3.   Domestic Crude Petroleum Reserves

From "Crude Petroleum — Domestic Reserves," *Chemical Economics Handbook*, p. 229. 3500, Stanford Research Institute. Data for 1973 are from *Oil and Gas Journal*, April 1, 1974, p. 42.

The general pattern is seen to be one of a decreasing life index. The projected lifetime of the reserves dropped below 10 years in 1968 and 1969. The inclusion of the Alaskan North Slope reserves increased the ratio in 1970, but it is now decreasing again and approaching 10 years.

Distribution of Reserves and Production

United States petroleum reserves as of January 1, 1974, were estimated at about 35.3 billion barrels. As might be expected, the reserve distribution is not uniform among the states. The reserves in Texas, Alaska, and Louisiana represent three fourths of the total — Texas reserves alone constitute one third of the national total. States with reserves of at least 1% of the national total are listed in Table 5—1. The values given for California, Louisiana, and Texas include adjacent offshore reserves.

As might be expected, a correlation exists between the location of petroleum reserves and the areas active in petroleum production. The correlation can be seen by comparing the reserve data of Table 5—1

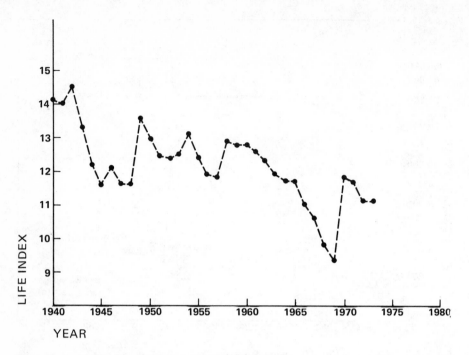

YEAR

Figure 5—4.   Life Index of Domestic Petroleum Reserves

Calculated from "Crude Petroleum – Domestic Reserves," *Chemical Economics Handbook*, p. 229-3500, Dec. 1972, Stanford Research Institute.

Table 5—1.   DOMESTIC PETROLEUM RESERVES BY STATE, JANUARY 1, 1974

State	Reserves (millions of barrels)	Percent of Total Reserves
Texas	11,756	33.3
Alaska	10,112	28.6
Louisiana	4,577	13.0
California	3,488	9.9
Oklahoma	1,271	3.6
Wyoming	917	2.6
New Mexico	643	1.8
Kansas	401	1.1
Other states	2,135	6.0
Total	35,300	100.0

Adapted with permission from "Another big drop reported for U.S. oil and gas reserves," *Oil and Gas Journal*, vol. 72, no. 13 (April 1, 1974): 42-43. Reprinted with permission from The Petroleum Publishing Company.

with the production figures given in Table 5—2. A comparison shows that the correlation breaks down in the case of two states, Alaska and Louisiana. Louisiana contains only 13.0% of the reserves but accounts for 22.6% of the production. Alaska, on the other hand, with 28.6% of the reserves accounts for only 2.3% of the production. The development of increased production from Alaska will be one of the important activities in the immediate future.

Figures 5—5 and 5—6 contain information about the worldwide distribution of petroleum reserves according to regions and countries. Nearly 56% of the estimated world reserves is located in one region, the Middle East. Most of the reserves within this region are found in four countries: Saudi Arabia, Kuwait, Iran, and Iraq. On a world basis, these four countries contain 45.8% of all petroleum reserves and rank first, third, fourth and sixth, respectively, in size of reserves. Eastern Europe's contribution to world reserves is located primarily in Russia (12.7%) while in Africa, the third largest reserve region, one country (Libya) contributes nearly half (4.1% out of 10.7%) of the regional total. The contributions of the North American region to world reserves comes mainly from the United States (5.5%) and Canada (1.5%). Of South America's 4.1%, 2.2% is found in Venezuela.

Chemical Composition of Petroleum

Crude petroleum is a complex mixture containing hundreds of different chemical compounds. Most of these compounds are struc-

Table 5—2.　ANNUAL DOMESTIC PETROLEUM PRODUCTION BY STATE, 1973

State	Production (millions of barrels)	Percent of Total Production
Texas	1,258	39.5
Louisiana	719	22.6
California	336	10.5
Oklahoma	180	5.7
Wyoming	138	4.3
New Mexico	96	3.0
Alaska	72	2.3
Kansas	67	2.1
Other states	319	10.0
Total	3,185	100.0

Adapted with permission from "Another big drop reported for U.S. oil and gas reserves," *Oil and Gas Journal*, vol. 72, no. 13 (April 1, 1974): 42-43. Reprinted with permission from The Petroleum Publishing Company.

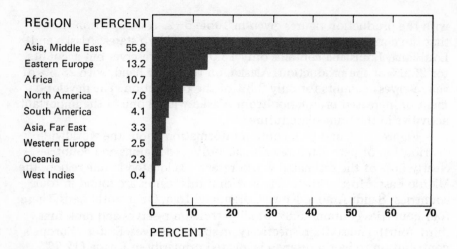

REGION	PERCENT
Asia, Middle East	55.8
Eastern Europe	13.2
Africa	10.7
North America	7.6
South America	4.1
Asia, Far East	3.3
Western Europe	2.5
Oceania	2.3
West Indies	0.4

PERCENT

Figure 5—5.   Regional Crude Petroleum Reserves, January 1, 1974

From "Crude Petroleum — World Reserves," *Chemical Economics Handbook*, p. 229.8000, Feb. 1973, Stanford Research Institute. Data for 1974 are from *Oil and Gas Journal*, Dec. 31, 1973, pp. 86-87.

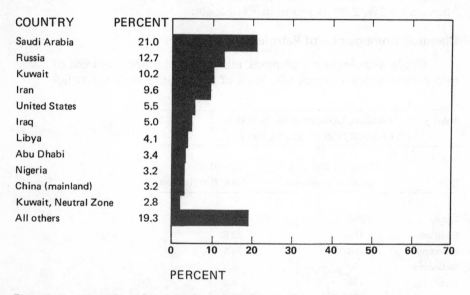

COUNTRY	PERCENT
Saudi Arabia	21.0
Russia	12.7
Kuwait	10.2
Iran	9.6
United States	5.5
Iraq	5.0
Libya	4.1
Abu Dhabi	3.4
Nigeria	3.2
China (mainland)	3.2
Kuwait, Neutral Zone	2.8
All others	19.3

PERCENT

Figure 5—6.   World Crude Petroleum Reserves by Country, January 1, 1974

From "Crude Petroleum — World Reserves," *Chemical Economics Handbook*, p. 229.8000, Feb. 1973, Stanford Research Institute (Russia added). Data for 1974 are from *Oil and Gas Journal*, Dec. 31, 1973, pp. 86—87.

turally similar to each other and are known as hydrocarbons. As the name suggests, hydrocarbons are composed of the two elements hydrogen and carbon. Small amounts of other elements have also been detected as a result of the careful analysis of crude petroleum samples.

The other elements found are nitrogen (0%—0.5%), sulfur (0%—6%), oxygen (0%—3.5%), and some metals in trace amounts. These additional elements are usually found incorporated into hydrocarbonlike molecules rather than in the free state.

Most petroleum hydrocarbons are composed of only three types of components — paraffin, cycloparaffin, and aromatic groups. Examples of these three types of components are shown in Figure 5—7.

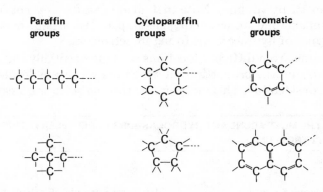

Figure 5—7.   Examples of Hydrocarbon Groups That Make Up Most Petroleum Hydrocarbons

Redrawn with permission from "Hydrocarbons in Petroleum" by F. D. Rossini, *Journal of Chemical Education,* vol. 37 (1960): Figure 1, p. 555.

In the groups of Figure 5—7, bonds (dashes) that do not connect two carbon atoms may be attached to hydrogen atoms or to other groups.

Three broad classes of hydrocarbons result from various combinations of hydrocarbon groups. These classes are:

1. Paraffin hydrocarbons — containing molecules composed only of paraffin groups.
2. Cycloparaffin hydrocarbons — containing molecules composed of one or more cycloparaffin groups that may or may not be attached to paraffin groups.
3. Aromatic hydrocarbons — containing molecules composed of one or more aromatic groups that may or may not be attached to cycloparaffin or paraffin groups.

The large number of molecular hydrocarbons found in petroleum makes the task of separating and identifying the various constituents quite difficult. In fact, no crude petroleum sample has yet been completely analyzed. The most extensive analyses have been carried out in work associated with the American Petroleum Institute. A mid-continent crude petroleum from Ponca City, Oklahoma, was selected as being representative of United States petroleum. As a result of using the best separation and identification techniques available, more than one hundred individual hydrocarbons have been isolated from the sample and identified. These isolated components represent only about one half of the original sample material and are mainly lower boiling hydrocarbons. Efforts to identify the higher boiling components continue. Of 175 hydrocarbons isolated, 70 are classified as paraffins, 48 are cycloparaffins, and 57 are aromatics. Table 5—3 contains a list of those compounds that occur in greatest abundance (greater than 1% by volume). Note that all are paraffins except for one cycloparaffin and one aromatic. In general, paraffins are the most abundant types of hydrocarbons found in petroleum.

Petroleums from different locations contain essentially the same hydrocarbons, but the proportions in which the various molecules occur vary considerably. This fact is not surprising since the local

Table 5—3.   THE MOST ABUNDANT HYDROCARBONS IN REPRESENTATIVE U.S. PETROLEUM

Formula	Name	Structural Formula	Amount (Volume Percent)
C_7H_{16}	*n*-heptane	$CH_3(-CH_2-)_5CH_3$	2.3
C_8H_{18}	*n*-octane	$CH_3(-CH_2-)_6CH_3$	1.9
C_6H_{14}	*n*-hexane	$CH_3(-CH_2-)_4CH_3$	1.8
C_9H_{20}	*n*-nonane	$CH_3(-CH_2-)_7CH_3$	1.8
$C_{10}H_{22}$	*n*-decane	$CH_3(-CH_2-)_8CH_3$	1.8
C_7H_{14}	methylcyclohexane		1.6
$C_{11}H_{24}$	*n*-undecane	$CH_3(-CH_2-)_9CH_3$	1.6
$C_{12}H_{26}$	*n*-dodecane	$CH_3(-CH_2-)_{10}CH_3$	1.4
$C_{13}H_{28}$	*n*-tridecane	$CH_3(-CH_2-)_{11}CH_3$	1.2
$C_{10}H_{14}$	1-methyl-3-isopropylbenzene		1.08
$C_{14}H_{30}$	*n*-tetradecane	$CH_3(-CH_2-)_{12}CH_3$	1.0

Adapted with permission from "Hydrocarbons in Petroleum" by F. D. Rossini, *Journal of Chemical Education*, vol. 37 (1960): Table 4, p. 558.

distribution of plant and animal life probably varied from region to region at the time petroleum was being formed just as it varies today. Differences in the percentage of the various hydrocarbons influence the physical properties of petroleum. Consequently, the properties of crude petroleums are extremely diverse. Some petroleums are almost colorless while others are pitch black, amber, brown, or green. *Ethereal* and *pleasant* are terms used to describe the odor of some petroleums. Others smell sweet or like turpentine or camphor. Still others have very unpleasant odors usually caused by the presence of certain sulfur-containing compounds.

Petroleum Refining

In its natural state, crude petroleum has very few uses. However, this complex mixture of hydrocarbons can be separated into various useful fractions by the process known as refining. The various resulting fractions are still hydrocarbon mixtures but each one is simpler (contains fewer components) than the original crude petroleum. The refining process consists of three types of processes: physical separation, alteration or conversion, and purification or treatment. The extent to which these processes are used is governed by the characteristics of the crude petroleum feedstock and the types of consumer products in demand.

Physical Separation

Physical separation is usually accomplished by fractional distillation, a process which makes use of the different volatilities of the various hydrocarbons. The volatility of a compound is related to its boiling point — the higher the volatility, the lower the boiling point. Hydrocarbons containing large numbers of carbon atoms have lower volatilities and higher boiling points than hydrocarbons composed of fewer carbon atoms.

Fractional distillation is accomplished by first heating crude petroleum to a temperature of $315°-370°C$ ($600°-700°F$). The resulting mixture of hot vapors and liquid is then passed into the bottom of a vertical fractionation tower or column. The temperature inside the tower decreases from the bottom to the top. Thus, as the vapors rise, they condense to liquids at different levels inside the tower. Those with low volatilities and high boiling points condense at lower levels while the more volatile components condense at higher levels. Each fraction is drawn off and stored until it is used or treated further. The process is represented by Figure 5—8.

The various fractions obtained from a typical fractionation process are tabulated in Table 5—4. The boiling points and some typical uses for each fraction are also included. In addition to uses such as those

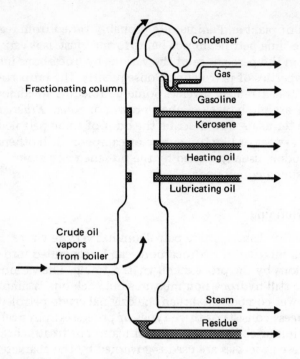

Figure 5—8.   Fractional Distillation of Petroleum

From *Chemistry: A Science for Today* by Spencer L. Seager and H. Stephen Stoker, p. 298.
Copyright © 1973 by Scott, Foresman and Company. Reprinted by permission of the publisher.

Table 5—4.   PRODUCTS OF PETROLEUM DISTILLATION

Fraction	Molecular Size Range	Boiling-point Range (°C)	Typical Uses
Gas	$C_1 - C_5$	−164 to 30	Gaseous fuel
Petroleum ether	$C_5 - C_7$	30 to 90	Solvent, dry cleaning
Straight-run gasoline	$C_5 - C_{12}$	30 to 200	Motor fuel
Kerosene	$C_{12} - C_{16}$	175 to 275	Fuel for stoves, diesel and jet engines
Gas oil or fuel oil	$C_{15} - C_{18}$	Up to 375	Furnace oil
Lubricating oils	$C_{16} - C_{20}$	350 and up	Lubrication
Greases	$C_{18} -$ up	Semisolid	Lubrication
Paraffin (wax)	$C_{20} -$ up	Melts at 52–57	Candles
Pitch and tar	high	Residue in boiler	Roofing, paving

From *Chemistry: A Science for Today* by Spencer L. Seager and H. Stephen Stoker,
p. 299. Copyright © 1973 by Scott, Foresman and Company. Reprinted by permission
of the publisher.

found in Table 5—4, a significant amount of the petroleum fractions produced annually is used as feedstock for the petrochemical industry. From this industry flow many consumer products that are considered to be necessities today. Examples of these products are plastics, synthetic fibers, adhesives, dyes, and medicinals.

Conversion Processes

The relative amount of each fraction obtained from fractional distillation depends on the composition of the original crude petroleum and usually does not correspond to the amounts needed to satisfy consumer demands. Chemical alteration or conversion enables refiners to change molecules found in the fractions with small demand into molecules corresponding to fractions in high demand. For example, only a small amount of gasoline, the fraction in highest demand, can be distilled from an average crude petroleum. Appropriate conversion processes make it possible to produce gasoline from hydrocarbons not found in the gasoline fraction. Three basic conversion processes are used to increase the yield of gasoline from crude petroleum: thermal cracking, catalytic cracking, and polymerization.

During thermal cracking large molecules, usually included in the heating oil fraction, are subjected to high temperatures and pressures. Under such conditions, some molecular bonds break and smaller molecules result. Some of these smaller molecules are of the sizes found in the gasoline fraction. A representative thermal cracking reaction is:

$$C_{14}H_{30} \xrightarrow[\text{pressure}]{\text{heat}} C_8H_{18} + C_6H_{12} \qquad \text{5--1}$$

The same sort of process is carried out during catalytic cracking. However, the reactions take place at lower temperatures and pressures because a catalyst is added. A catalyst is a substance that accelerates chemical processes without itself undergoing any change. Various acids, clays, and metals such as aluminum and platinum are used as catalysts for cracking reactions.

Polymerization is the reverse of cracking. It is a conversion process in which smaller molecules are combined to form larger ones. Molecules smaller and more volatile than gasoline are formed into larger, gasoline-type molecules as represented by the following example:

$$C_4H_8 + C_3H_6 \longrightarrow C_7H_{14} \qquad \text{5--2}$$

Petroleum Purification

The third type of commonly used refining process is purification. One purification step of special importance is the removal of sulfur-containing compounds from refinery products. Sulfur-containing compounds are undesirable for several reasons:

1. Some impart unpleasant odors to petroleum products.
2. Some are corrosive to refinery equipment, storage facilities, and engines.
3. They "poison" or deactivate catalysts used in refining processes.
4. They react with lead antiknock gasoline additives and reduce their effectiveness.
5. They form sulfur oxide air pollutants when petroleum products are burned.

Most of the older purification treatments were aimed at eliminating objectionable odors caused by sulfur-containing compounds. These treatments usually did not remove the guilty compounds but merely converted them into less odorous types. For example, "copper sweetening," a process represented by equation 5—3, was used to eliminate the odor caused by mercaptans (butyl mercaptan, C_4H_9SH, is responsible for the characteristic odor of skunks). In reaction 5—3, R stands for a hydrocarbon group. The resulting disulfides have much less objectionable odors than the mercaptans.

$$2RSH + 2CuCl_2 \longrightarrow R_2S_2 + 2CuCl + 2HCl \qquad 5\text{--}3$$

In recent years the actual removal of sulfur compounds (desulfurization) has become increasingly desirable. One reason for this is to preserve the activity of refining catalysts. Another is to enable the petroleum products to meet air pollution standards upon combustion. Sulfur removal is usually accomplished by a hydrogen treatment in which hydrogen gas reacts with the sulfur of objectionable compounds to form the easily removed hydrogen sulfide gas. A typical reaction is:

$$C_4H_4S + 4H_2 \xrightarrow[\text{heat, catalyst}]{\text{pressure}} C_4H_{10} + H_2S \qquad 5\text{--}4$$

Hydrogen treatment is carried out at moderate temperatures and pressures. Oxides of cobalt and molybdenum serve as catalysts. In addition to removing sulfur compounds, the hydrogen treatment improves the stability of petroleum products — the replacement of sulfur by hydrogen creates hydrocarbon mixtures of high purity with a strong resistance to deterioration.

Modern Petroleum Refining

Refinery processing has developed to the extent that today's refinery output is continually altered in order to adjust to seasonal and regional consumer demands. During winters in cold climates when the demand for furnace oil is high, its output is increased. During summer months when the demand for gasoline is great, its output is maximized at the expense of fractions in less demand.

The relative importance of the various fractions obtainable from crude petroleum has changed dramatically with time. Kerosene for use in lamps and stoves was in greatest demand until about 1900. Fuel oils and lubricants were prepared in smaller amounts, and gasoline was an undesirable side product which created disposal problems. The automobile then entered the scene and the petroleum industry was provided with a growing outlet for a previously useless by-product. Today, gasoline is the major product of petroleum refining. Figure 5—9 shows the consumption levels of the various petroleum fractions for the period since 1940.

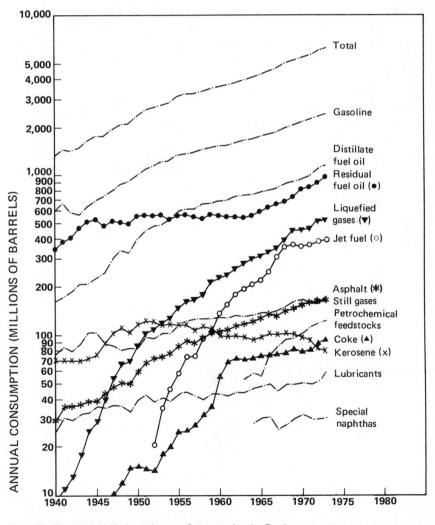

Figure 5—9.   Liquid Hydrocarbons — Consumption by Product

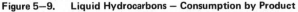

From "Primary Liquid Hydrocarbons — Consumption by Product," *Chemical Economics Handbook*, p. 229.3020A, Jan. 1974, Stanford Research Institute.

Distillate fuel oils are seen to be second in demand to gasoline. The main use of these oils is the comfort heating of homes and small apartment houses and buildings. Industry is the major consumer of residual fuel oil which is used to fire open hearth furnaces, steam boilers, and kilns. Large apartment and commercial buildings rank second as residual fuel oil consumers.

Problems Related to Petroleum Use

Petroleum demands in the United States exceed domestic production; as a result, some needs are met by imports. This situation results from many factors of diverse origin which continue to operate and cause the gap between demand and domestic supply to steadily increase. This increasing dependence on imports is the main problem associated with petroleum use in the United States. Figure 5—10 shows

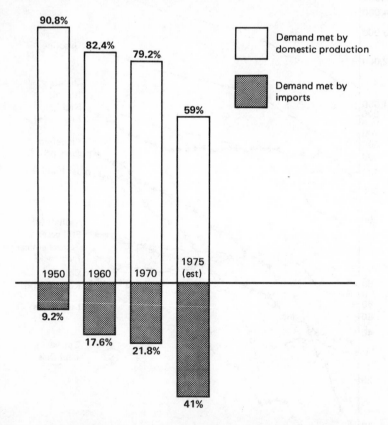

Figure 5—10.   Increasing Dependence of the United States on Petroleum Imports

Data from "Trends in Oil and Gas Exploration," Part 1, U.S. Senate Subcommittee Hearings, Aug. 8—9, 1972, p. 343 and "Relationship of Energy and Fuel Shortages to the Nation's Internal Development, House of Representatives Subcommittee Hearings, Aug. 1—3, 8—11, 1972, p. 961.

how the dependence on imports has increased since 1950.

The factors causing an increased dependence on imports and those problems resulting from this increased dependence will be discussed by considering the following topics:

1. Domestic Supply-Demand Imbalance
2. Petroleum Import Sources
3. Risks Accompanying Import Dependence
4. Petroleum Transportation Problems
5. Decreased Exploration for New Petroleum Fields

Domestic Supply-Demand Imbalance

Simply stated, the increasing supply-demand gap for domestic petroleum is the result of a demand that is growing more rapidly than expected and a domestic production that has peaked (see Figure 5—3) and now remains essentially static. Thus, the causes of the problem can be divided into two groups — those factors causing unexpected growth in demand and those resulting in a leveling of production.

The unexpected growth in petroleum demand is the result of several factors. Environmental quality has become an important issue in the United States; the impact of pollution control rules and regulations on the use of various fuels has been significant. Air pollution standards create a need for low-sulfur fuels. Shortages of natural gas (see Chapter 6) and the limited availability of low-sulfur coal (see Chapter 7) have resulted in an increased use of petroleum-based fuels. Another environmental protection decision, the use of automotive emission controls, has had a profound effect on petroleum consumption. It has been estimated that automobiles produced in 1972 required 300,000 more barrels of petroleum per day (about 110 million barrels per year) to travel the same number of miles as the automobiles they replaced. Emission controls decrease pollutant output but most of them also lower engine efficiencies.

Nuclear sources were expected to satisfy a significant part of the increasing energy demands of the '70s. However, a unique set of environmental problems accompanied the development of nuclear-generating facilities (see Chapter 8). As a result, nuclear energy supplies have been developed far slower than was anticipated; petroleum fuels were again used to make up the difference.

At the present time, domestic petroleum production facilities are operating at nearly maximum capacity. This has not always been the case. During the twenty years preceding 1970, the petroleum industry had surplus production capacity that was used to satisfy regular growing demands and short-term or emergency requirements for petroleum products. This surplus capacity no longer exists. It has been pressed into regular use to satisfy petroleum demands caused, at least in part, by the factors previously discussed. New production capacity has not been developed to keep up with demands and, as a result, petroleum production has leveled off.

The failure to increase petroleum production facilities has been caused partially by local, state, and regional concern for the environment. The construction of refineries and pipelines, and some exploration activities have gone slowly because of the imposition or concern about the possible imposition of environmentally related restrictions. The controversy surrounding the Alaskan pipeline construction is a well-known example.

Influences other than environmental concern have also contributed to the present production capacity shortage. Many of these influences result from trends in the economic and political climate of the United States. Most of these trends were set in motion years ago and include the following:

1.   Significant amounts of the nation's potential supply of energy — including offshore petroleum reserves and oil shale deposits — are located on federally owned land. Leasing practices in the past have not allowed such resources to be developed at the pace needed to keep up with demand.

2.   Money used by the petroleum industry for expansion and exploration comes from profits. Thus, a reasonable expectation of economic reward must be present before industry will invest in resource development. Various pressures, including the possible elimination of the oil imports control program and the tax reform act of 1969, caused the economic climate for investments to deteriorate steadily during the past decade.

3.   An atmosphere of uncertainty pertaining to investments in domestic refining facilities was created during the early 1970s. This came about because of policies in the oil imports program. Numerous special exemptions and temporary control suspensions created uncertainties about supplies of imported petroleum and petroleum products. Simply stated, investments in new refineries were held back until it was determined whether or not the refined products would have to compete against excessive foreign imports. Another concern was whether or not foreign crude petroleum would be available as feedstock for the contemplated refining facilities.

If we assume discoverable petroleum to be represented by cumulative production, proven reserves, and probable reserves, the data of Figure 5—2 indicate that more than 80% of the discoverable domestic petroleum is still to be produced. Thus, the problem of static production is not caused by a lack of potential domestic petroleum supplies but by a lack of exploratory activity and refinery construction.

Petroleum Import Sources

Answers to the following three questions provide useful background information related to the problems associated with a United States dependence on imported petroleum:

1. What is the location of the world's petroleum reserves outside the United States?
2. What were the sources of petroleum that was imported into the United States in the recent past?
3. To what extent will the United States depend on petroleum imports in the future; what foreign nations will be involved?

The first question was discussed earlier and the answer is summarized in Figures 5—5 and 5—6.

Data indicating the sources of imported United States petroleum are given for selected years in Table 5—5. It is evident from these data that during the 1960s most United States imports came from the Western Hemisphere and especially from the Caribbean area. The two most important Caribbean suppliers were Venezuela and the Netherlands West Indies. The contribution of the Western Hemisphere to the total United States petroleum imports increased from 77% in 1960 to 84% in 1970.

Since 1970 United States import patterns have changed dramatically, as shown by Figure 5—11. An ever increasing demand for

Table 5—5.   SOURCES OF UNITED STATES IMPORTED CRUDE PETROLEUM AND PETROLEUM PRODUCTS
(millions of barrels)

	1960	1965	1970	1975 (est.)
Western Hemisphere				
Canada	44.5	117.5	280.0	474.5
Caribbean	446.0	561.7	631.1	565.8
Other countries	20.1	38.7	132.1	434.4
Eastern Hemisphere				
Middle East	95.3	108.4	67.5	983.7
Other countries	56.6	74.5	137.2	271.9
Total	662.5	900.8	1,247.9	2,730.3

Data from "Trends in Oil and Gas Exploration," Part 1, U.S. Senate Subcommittee Hearings, Aug. 8—9, 1972, pp. 343, 450.

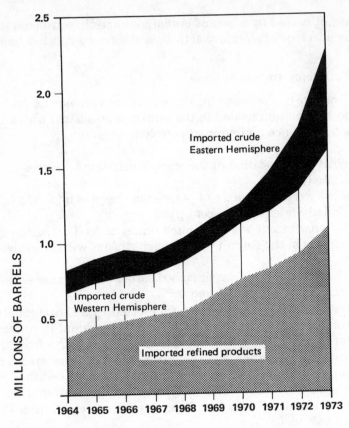

Figure 5—11.   United States Petroleum Imports

Drawn from data in *Oil and Gas Journal,* January 28, 1974, p. 119.

petroleum has more and more been satisfied by shipments from the Eastern Hemisphere. It is apparent from Figure 5—11 that crude petroleum is only one form in which petroleum is imported. In fact, it was not until 1973 that the quantity of imported crude petroleum exceeded the quantity of imported refined products. In addition, Figure 5—11 shows that 1973 was the year in which crude petroleum imports from the Eastern Hemisphere exceeded those from the Western Hemisphere.

Projections made before the 1973—74 Arab oil embargo indicated that the percent of United States petroleum demands satisfied by imports would continue to increase. These same projections revealed that most of the increase would come from the Eastern Hemisphere — the location of the majority of the world reserves (recall Figure 5—6). Thus, the trend shown in Figure 5—11 was expected to continue.

As a result of the embargo, a national energy policy (Project Independence) was established which seriously affects the accuracy of these projections. According to this policy, domestic energy sources will be intensively developed in an effort to make the United States energy

Table 5—6.   Estimates of Future United States Dependence on Imports of Crude Petroleum
and Petroleum Products

	Preembargo Estimate		Postembargo Estimate	
	1980	1985	1980	1985
Total estimated demand (billions of barrels/year)	8.1	9.4	6.7	6.9
Domestic Production	53%	41%	59%	71%
Imports	47%	59%	41%	29%

Preembargo estimates from "Trends in Oil and Gas Exploration," Part 1, U.S. Senate
Subcommittee Hearings, Aug. 8—9, 1972, p. 343. Postembargo estimates calculated from
data in *Oil and Gas Journal,* Aug. 19, 1974, p. 33.

self-sufficient by 1980. A great deal of skepticism has been expressed
about the possibility of attaining import independence as early as 1980.
A more realistic target date of the late 1980s has been suggested. Until
the goal is reached, petroleum imports will continue to be an important
part of U.S. energy supplies. Table 5—6 gives the preembargo estimate
of import dependence as well as a postembargo goal based on Project
Independence (which includes significant energy conservation
measures). Note that this estimate projects the United States to be
70% self-sufficient for petroleum supplies by 1985.

Risks Accompanying Import Dependence

The advisability of U.S. energy independence becomes apparent
when the actual and potential problems related to petroleum importation are recognized. We have seen that the Middle East has petroleum
reserves large enough to meet U.S. import needs for the foreseeable
future. The question, then, is not one of quantity but of availability.
On what terms would such petroleum be available to potential
customers? A number of factors could influence such terms, including
price, control of production and marketing facilities, and political
considerations.

By relying on imports to the extent projected by Table 5—6, the
United States would become vulnerable to two types of risk: (1) Those
involving the threat of either partial or complete supply interruption by
producing countries to force economic concessions and (2) the actual
interruption of supplies in time of war or in times of political turmoil
and tension short of war. The security of the United States when faced
with such risks depends on the degree of dependence on foreign
petroleum supplies and the ease with which alternative supplies can be
obtained and distributed.

Some specific problem areas associated with these risks are
identified and discussed below. It should be noted that some of the

problems have occurred in the past while others may or may not materialize, depending on numerous factors.

1. Uncertainties in time of war — During war or the tenuous times before war, the safety of petroleum transporting tankers would be in doubt on open seas. Their vulnerability to attack is well known. In addition, a war might possibly increase United States petroleum requirements beyond the ability or willingness of foreign sources to supply.
2. Governmental changes in producing countries — Governmental control in exporting countries might be achieved by political factions unwilling to do business with the United States. This could result in a cutoff of petroleum supplies with little or no forewarning. In addition, local or regional revolution, hostilities, or guerilla activities might interrupt production or transportation in exporting countries for extended periods of time even though no government changes occur.
3. Political blackmail — The reliance on a small group of nations for a large volume of petroleum imports always presents the possibility that they might act together and refuse to sell oil to the United States because of differences in political points of view. This was the cause of the 1973—74 embargo, when ten nations completely stopped petroleum exports to the United States because of continued U.S. support of Israel.
4. "Hand on tap" policy — Some observers believe that petroleum-producing nations, while not stopping production, might limit their production in the future. Such limitations could be prompted by a desire to protect and conserve their resources or an attempt to increase prices. The first reason discounts the size of the huge reserves that exist in the Middle East and the fact that production limitation carries with it the risk of losing markets to competitors. The second motive — to increase prices — could have some validity and provides incentives to help promote development in Western countries.
5. Attempts to force economic concessions — In an attempt to gain greater bargaining power in their dealings with international petroleum companies, a significant number of the major oil-exporting countries joined together in 1960 to form the Organization of Petroleum Exporting Countries (OPEC). In 1974 the member countries were Saudi Arabia, Iran, Kuwait, Iraq, Abu Dhabi, Quatar, Libya, Algeria, Nigeria, Indonesia, and Venezuela.
    A significant accomplishment of this alliance has been the establishment of petroleum prices at higher levels. A five-year agreement was reached in 1971 between the Persian Gulf countries and the United States. The result was a substantial price increase to $3.40 per barrel for Persian Gulf petroleum. The other OPEC

members followed this with equal or greater price increases to the United States. In the second year of this agreement, the OPEC countries negotiated for and received further price increases to a level of $3.80 per barrel as compensation for devaluation of the dollar. Between 1973 and January 1 of 1974, the price per barrel of OPEC crude petroleum was increased drastically to $10.30, and for the first time became higher than that of domestic crude ($7.60 per barrel). OPEC countries are also seeking to participate as part owners in the companies exploiting their resources. If the OPEC members maintain a united front despite their diverse national interests and historical differences, they could continue to exert great pressure for economic and political concessions from their petroleum customers. They might even again resort to a united interruption of petroleum supplies.

6.   Balance-of-payment problems — A heavy reliance on foreign petroleum supplies creates serious balance-of-payment problems which relate to the position of the United States in world economics. It is estimated that the trade deficit related to U.S. fuel and energy purchases (mostly petroleum) will increase from approximately $3.7 billion in 1970 to more than $20 billion by 1975 as the result of increased imports and drastic price increases. In worldwide balance-of-payment terms, the impact of a trade deficit of this size could be very significant. It can be lessened by the earnings from increased exports to petroleum-producing countries, the earnings of tanker fleets, and increased investments in the United States by petroleum-producing countries. However, it is difficult to foresee the manner in which the total of such a large deficit could be accommodated.

7.   Competition for petroleum supplies — Both western Europe and Japan depend on petroleum imports to a greater extent than the United States. In 1973, they imported 60% and 75%, respectively, of the total they required. These heavily dependent nations are concerned about the unexpected rate at which United States imports have increased. Their concern, of course, is related to the impact U.S. imports will have on their supplies. American companies dominate the international petroleum industry, even though some have been nationalized or are being divested. This fact creates fears that American needs will be satisfied first.

Keen competition between Japan, western Europe, and the United States for petroleum supplies could significantly affect prices. The nature of the bargaining process depends on the number of consumers and their demands. Producer countries could play the consumers against one another by holding down production. The significance of this problem can be seen from Figure 5—12, which shows the world petroleum consumption and major movements of supplies for 1970 and the preembargo

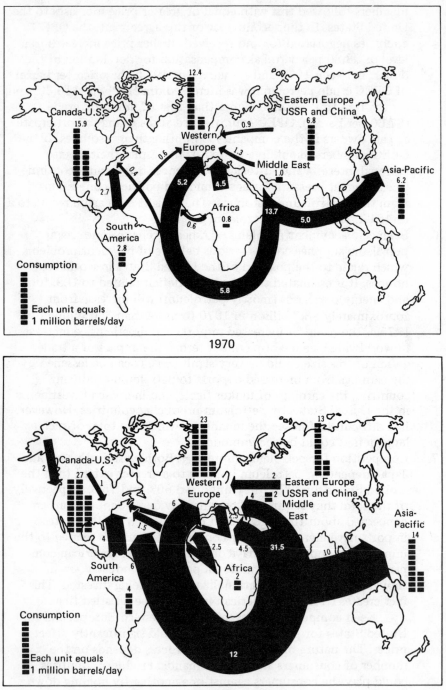

Figure 5-12.   World Petroleum Consumption and Major Movements

From *The National Energy Outlook*, Shell Oil Company, March 1973, pp. 32-33.

projection for 1980. Note that only a small amount of Middle East petroleum came to the United States in 1970. Although this amount is projected to increase substantially by 1980, it will still be smaller than that going to western Europe and Japan.

Petroleum Transportation Problems

In addition to the previously discussed problems related to economics and military security, the use of petroleum imports also raises another question: What is the most expedient way to transport this huge volume of petroleum to the United States? It must be done as cheaply and with as little damage to the environment as possible. Studies conducted by the Council on Environmental Quality indicate that less pollution will result from the use of fewer but larger tankers and deepwater port facilities than from the use of the many small tankers and conventional facilities that would be needed. The use of larger ships also cuts transportation costs. These conclusions lead to the problem of providing supertankers and deepwater ports.

World bulk shipping is undergoing drastic changes in the size of oceangoing vessels used to transport petroleum and other bulk materials such as coal and iron ore. In 1950 most bulk materials were carried in 16,000-ton capacity tankers; the largest bulk carrier had a capacity of 24,500 tons. The emergence of the 100,000-ton and larger bulk carrier has taken place since the mid-1960s. As recently as 1966, there was only one ship in the world capable of carrying over 200,000 tons. By 1974 the world fleet of bulk vessels with capacities in excess of 100,000 tons numbered about 800, of which 400 had capacities in excess of 200,000 tons. By 1980 the 200,000- to 300,000-ton capacity tanker and bulk carrier will be the standard vessels used to carry large quantities of bulk materials.

Significant freight cost savings are made possible by such vessels. The dollar-per-ton cost can be reduced by nearly 30% when goods are shipped in 250,000-ton tankers rather than 65,000-ton tankers. Even greater savings can be realized by using still larger ships. These highly productive and economical vessels require no more time in port than smaller ships because of faster loading and discharge equipment. If the United States fails to take advantage of such vessels, it will have to pay the resulting higher transport costs for petroleum and petroleum products. In addition, a competitive edge will be lost because of increased delivery costs for United States coal exported to foreign markets.

In order to accommodate these mammoth tankers, ports with water depths of from 60—90 feet are required. The present U.S. port system, particularly on the North Atlantic where traffic is the greatest, is completely inadequate for such use. With a few exceptions on the West Coast, existing channel depths of 35—45 feet at most U.S. ports limit the size of bulk vessels served to about 80,000 tons capacity. The challenge of the supertankers is being met impressively by foreign ports; as of early 1974, over 50 foreign deepwater port facilities were in operation, under construction, or planned with capabilities for handling 200,000-ton or larger ships. Europe and Japan are increasingly being supplied raw materials by supertankers.

The most economical alternative to providing deepwater port facilities on the Atlantic coast is an offshore transfer terminal located in a natural deepwater harbor. Such facilities have the environmental advantage of reducing the risks of collisions or groundings and the subsequent damage from spills. This advantage is gained by locating the facilities away from estuaries, rivers, and channels used by smaller vessels.

Without deepwater ports or offshore terminals, both U.S. and foreign companies will have to expand the use of presently operating petroleum transshipment terminals located in the Bahamas and the Canadian maritime provinces. At these terminals, petroleum is transferred from large vessels to smaller ones which then use conventional port facilities. The use of conventional ports will increase and will thereby increase the risk of pollution from shipping operations and accidents. At the same time, the United States will lose the jobs and capital associated with the work done at the foreign transshipment terminals.

Decreased Exploration for New Petroleum Fields

The trends in petroleum exploration activities from 1946 to 1970 are shown in Figure 5—13. Exploratory wells are those drilled for three purposes: (1) to find the limits of a petroleum-bearing formation that is partly developed, (2) to search for new deposits in an area that is already productive, and (3) to locate deposits in areas where neither petroleum nor natural gas has ever been found. The last of these three types of exploratory wells is known as a new-field wildcat well. The geophysical crew-months worked gives a measure of the effort expended for the purpose of locating prospective drilling areas using geophysical methods.

The United States emerged from World War II with a rapidly increasing requirement for petroleum and a production capability that had been curtailed by wartime restrictions. As a result, exploratory activities rapidly increased and finally peaked in 1957. Petroleum demand in the U.S. then slowed to a lower growth rate and foreign supplies began to overload the market. Under such conditions, domestic production increases were not needed and a readjustment period followed during which geophysical exploratory work and drilling were curtailed. Petroleum demand again began to grow at an accelerated rate by 1967, but drilling and geophysical exploration did not show a

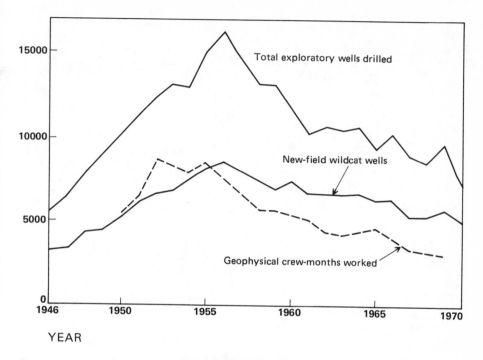

Figure 5—13. United States Petroleum Exploratory Activities

From *U.S. Energy: A Summary Review*, Dept. of the Interior (Washington, D.C.: Government Printing Office, January 1972), Figure 9, p. 28.

corresponding increase. Instead, as shown in Figure 5—13, they remained at relatively low levels.

The lack of interest in exploration and other developmental activities in the late 1960s has been attributed to a number of factors. The factors, all of which add up to insufficient economic incentives, are:

1. A long-term decline in the real price of domestic crude petroleum. This is shown in Figure 5—14 in terms of constant 1958 dollars. Note that the real price for crude petroleum dropped steadily from 1957 to 1971. Only since 1973 has it increased.
2. Uncertainties surrounded federal policy concerning petroleum import controls.
3. It became increasingly difficult to find new petroleum and natural gas deposits that were large enough to permit economic production. This factor is perhaps the most significant. Figure 5—15 shows the diminishing success of exploratory efforts as indicated by the number of barrels of petroleum found per foot of exploratory well drilled. This chart probably understates recent discoveries

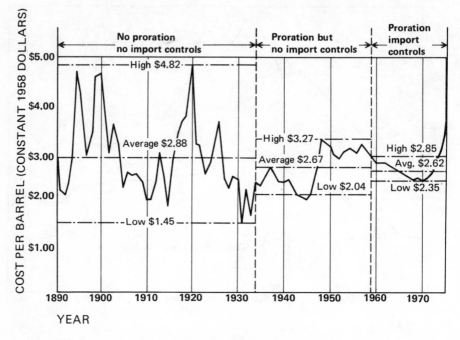

Figure 5—14.   The Price of United States Crude Petroleum,

Redrawn from "Trends in Oil and Gas Exploration," Part 1, U.S. Senate Subcommittee
Hearings, Aug. 8—9, 1972, p. 185.

since the full size of new fields typically is not identified until
8—10 years after their discovery.

A conclusion similar to that implied by the data of Figure
5—15 is obtained from a somewhat different viewpoint — the
number of new wildcat wells required to find one significant
deposit of petroleum or gas. A significant deposit is one which
ultimately yields more than one million barrels of crude petroleum. Only about thirty wildcat wells were needed to find a
significant deposit in the late 1940s; the number of wells required
had nearly doubled by 1960, and the trend has not been changed.

An important internal situation helped kindle renewed interest in
petroleum exploration. As economic incentives decreased, it was often
more economical for the petroleum industry to provide additional
reserves and production capacity by intensively developing known
reservoirs rather than exploring for new ones. Reservoir engineering
became more sophisticated and efficient, and the recovery percentage
of petroleum from known reservoirs increased. However, as the 1960s
drew to a close, it was becoming apparent that the number of opportunities for the application of improved recovery technology was

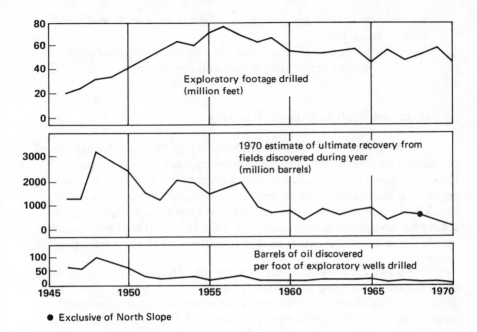

• Exclusive of North Slope

Figure 5—15.   United States Petroleum Exploration Effort and Results

From *U.S. Energy: A Summary Review,* Dept. of the Interior (Washington, D.C.: Government Printing Office, January 1972), Figure 10, p. 28.

decreasing — the technology had been used on most of the suitable reservoirs. Therefore, the future of the petroleum industry would depend heavily on new petroleum discoveries.

Methods for Increasing Petroleum Production

The remainder of this chapter is devoted to a discussion of topics which represent possible ways to increase domestic petroleum supplies in the future. Such action is necessary if the desirable goal of reducing import dependence is to be reached. Increased domestic production requires the construction of new refineries and, in addition, an increase in the amount of domestic crude petroleum available for refining. A number of activities will contribute to an increase in the production of domestic crude petroleum:

1. Increase the rate of exploration for and discovery of new petroleum fields.
2. Increase the recovery percentages to provide more production from existing resources by a greater use of primary and secondary recovery techniques.
3. Develop or improve the technology for obtaining petroleum from

nonconventional sources such as oil shale, tar sands, and waste materials.
4. Develop the outer continental shelf, the most probable area for the discovery of major new petroleum supplies.

Increased Exploration and Discovery

The response of the petroleum industry to the need for increased petroleum reserves and production capacity has been affected in important ways by environmental considerations. In order to meet regulatory standards, more money must be spent during exploration, development, and production activities. Without disputing the desirability of environmental standards, it must be noted that the resulting increased costs, unless compensated for, diminish the attractiveness of investment.

The downward trend in exploration and discovery will be reversed when such activities become more economically attractive. Significant progress in this direction will depend on changes in governmental policies related to the petroleum industry and technological developments that decrease exploration risks.

It is the opinion of some that the first requisite for finding more petroleum and natural gas is access to public land resources located both onshore and offshore. So many of the potential resources lie under public lands that the leasing policies of the government assume major importance. The tax treatment of exploratory and producing operations will also significantly influence the amount of exploration done.

The current success ratio (number of discoveries per number of wells drilled) for new significant petroleum deposits is about 1:60. It is hoped that advances in geophysical and geological techniques can lead to a greater ability to pinpoint suitable drilling targets. However, the wells must still be drilled in order to determine whether or not petroleum is present. Therefore, improved developmental technologies such as drilling and reservoir evaluation which lead to reduced costs would also provide economic incentives for exploration and development.

An example of technological improvements is the seismic reflection technique used to successfully map subsurface structures and locate potential drilling sites. The technique involves the production of a shock wave by mechanical devices or explosives. The shock waves travel downward where they strike rock formations. The waves are reflected from the rock formations back to the surface where they are detected electronically and recorded. The technique is represented diagrammatically in Figure 5—16. The geophysicist determines the general characteristics of the underground structure by correlating the

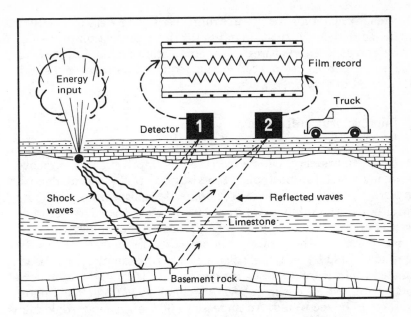

Figure 5—16.   The Seismic Reflection Technique for Subsurface Mapping

Redrawn with permission from "Exploration" in *Facts About Oil*, 1960 ed., p. 3, published by the American Petroleum Institute.

intensity of the reflected waves and the time interval required for them to travel to and from the rock formations. During the last decade, large amounts of research money have been spent to improve the seismic technique. Some important developments include the increased use of nonexplosive shock-wave sources, computer processing of field data, and computer-assisted interpretation of the collected data.

Most shallow, easily found, domestic petroleum deposits have already been developed. Because of this, most domestic exploration and development in the future will necessarily be directed toward:

1. Very deep deposits — located at depths where pressure and temperature are high.
2. Deepwater deposits located offshore.
3. Deposits located under permanent ice.
4. Deposits located under ice-laden and/or stormy seas.
5. Deposits contained in poor quality reservoirs.
6. Deposits located under difficult-to-develop terrain.

Some technological obstacles exist which must be overcome in order for work to progress in these areas. For example, there are no

high-strength materials for use in drilling very deep wells and only
insufficient design information exists for the construction of very
deepwater offshore structures. Also, completion techniques (borehole
lining, tower construction, etc.) are inadequate for deeper wells and
wells located in very deep water.

Improvement of Recovery
Efficiencies

Petroleum reserves can be increased by improvements in recovery
technology which can be applied to existing as well as future deposits.
The average recovery efficiency for domestic wells is presently about
31%, although the value varies from 10% to more than 75% for individual deposits. Predictions are that improved technology can
ultimately increase the average to about 60%. An increase of this
magnitude would be extremely significant. For example, it has been
estimated that future domestic discoveries will total 436 billion barrels
of petroleum. With existing recovery technology, 31% or 135 billion
barrels could be recovered. An increase to 60% efficiency would allow
the recovery of an additional 126 billion barrels — a 24-year supply at
1973 consumption levels. It is obvious that recovery technology is an
important area for intensive research and development activities.

The recovery of petroleum from a well involves primarily a
displacement process. The petroleum must be displaced out of the
reservoir rock formation and allowed to collect in the bore of the well.
Two natural displacing agents, natural gas and water, are often found
along with petroleum (see Figure 5—1). The variations in pressure and
displacing ability of these materials contribute significantly to the
production characteristics or phases of a well. These stages in a well's
productive life are called the flush, settled, and stripper phases.

Flush production is usually, though not always, the first phase in a
well's productive life. Flush production occurs when sufficient natural
pressure is available to force the petroleum to the surface.

Settled production is reached when the initial pressure dissipates
somewhat and the production rate tapers off (settles) to a fairly
constant daily average. In this phase, the displacement pressure is
usually too low to force the petroleum to the surface, and pumps must
be used. However, the displacement pressure must still be high enough
to force the petroleum out of the reservoir rock and into the well bore.
Some wells never flow naturally and must be pumped from the beginning while others drop sharply in flow rate shortly after production
begins and become settled fairly early.

The *stripper* or marginal phase of production is reached when the
output of a well drops below its settled rate. Stripper wells are usually
older and produce less than 10 barrels per day. They are kept in
production because output is steady and the yield is good over a long

period of time. They are generally pumped only intermittently to allow the petroleum to accumulate in the well bore. There are now approximately 350,000 stripper wells in this country; their slow but steady output provides about 13% of the domestic petroleum production.

Secondary Recovery Techniques

The natural flow and pump-assisted recovery techniques are known as primary recovery methods. They fall far short of accomplishing total recovery, but in the early years of the petroleum industry they were the only techniques available. During that time pumping was stopped and wells were capped when natural pressure could no longer force petroleum out of the reservoir rock into the well bore. As much as 70%—80% of the petroleum was left in the ground by such practices.

Today, secondary recovery techniques are used which prolong the productive life of a well. In one approach this is accomplished by a repressurizing technique in which gas or water is injected into the reservoir rock formation. In both methods, carefully located auxiliary service wells are drilled to create an efficient injection pattern. Water and gas injection techniques are represented in Figure 5—17. When a petroleum field becomes depleted to the point that repressurizing techniques are no longer effective, other secondary recovery methods are used.

One of these methods, *water flooding*, makes use of water injection. However, instead of injecting just enough to restore the natural displacement pressure, water is used in large enough quantities to create a water front which moves through the reservoir rock and pushes the petroleum ahead of it. In an attempt to further improve this

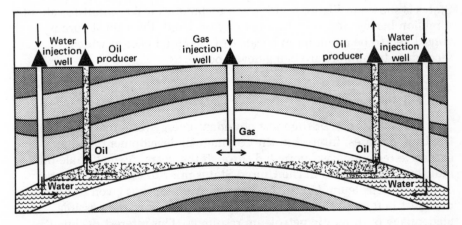

Figure 5—17.   Water and Gas Injection Recovery Techniques

Redrawn with permission from "Water and Gas Injection" in *Facts About Oil,* 1971 ed., chapter 5, p. 17, published by the American Petroleum Institute.

process, detergents are sometimes added to the water to keep wells from plugging up and corroding. In some cases carbon dioxide gas is dissolved in the water. The gas, in turn, dissolves in the petroleum, reduces its viscosity, and allows it to flow more easily.

Thermal recovery methods are also used effectively to increase the productivity and lifetime of wells. *Steam injection* is the newest variation in thermal techniques and currently finds wide use. In this technique, steam is injected into a producing well over a period of 10 days to 2 weeks. The well is shut down for a week or so to allow the reservoir to become thoroughly heated. The heat from the steam is locked in the porous rock formation and reduces the viscosity of the contained petroleum. The petroleum then flows out of the reservoir rock more readily.

Fire flooding, a second thermal recovery technique, makes use of heat energy provided by burning some of the petroleum still in the reservoir rock. Air or oxygen-containing gas is injected into the reservoir. The petroleum-air mixture is then ignited. The injected air is converted by combustion into a hot stream of gases which heats the petroleum, expands it, and reduces its viscosity. The combustion front slowly advances through the formation, often moving only 3 inches per day. The combustion more closely resembles a glowing charcoal briquet than the blaze of a campfire. Most of the heat is stored in the rock upstream from the flame. As a result, a high-temperature region passes through the reservoir rock and pushes the petroleum ahead of it, as illustrated in Figure 5—18. At first glance this technique appears to be wasteful. However, it really isn't; every recovery process leaves some petroleum behind, and it is reasonable to burn a portion of the non-recoverable petroleum in order to increase the amount that is recovered.

The significance of secondary petroleum recovery in the United States is shown in Figure 5—19. Such techniques will account for about 4 million barrels of production per day by 1980. Even this figure is considered to be a minimum because of the rapid pace at which secondary techniques are being developed.

Oil Shale

Oil shale is one of the most abundant energy resources found in the United States. Interest in the commercial development of this extensive resource has fluctuated through the years. A small shale-oil industry was operated in the United States prior to the 1859 discovery of natural petroleum. No further serious attention was given to oil shale until just before 1920 when concern developed about the possible inadequacy of domestic petroleum resources. This interest declined with the discovery and development of ample deposits of liquid petroleum. Only intermittent shale-oil projects have been conducted

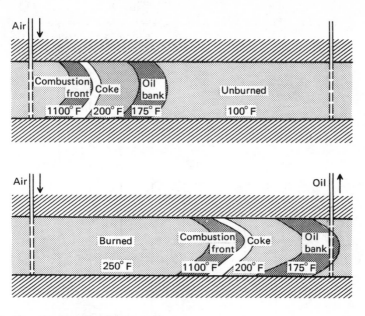

Figure 5—18.   The Fire-Flooding Technique

since then, first by the Bureau of Mines and more recently by commercial petroleum companies. No commercial petroleum production has resulted, but technologies have been developed which are considered to be applicable to commercial operations. As petroleum demand and prices increase, more and more attention is being given to this potential petroleum source.

Oil shale, a very fine-grained sedimentary rock, does not contain liquid petroleum. Rather, it contains a solid, largely insoluble organic material called *kerogen*. This substance is found associated with a mixture of inorganic minerals that make up the bulk of the rock.

The chemical difference between petroleum and kerogen is primarily one of molecular geometry. Petroleum molecules are typically composed of linear chains with some attached rings and branches (see Figure 5—7). Very little linking is found between the chains. In kerogen, on the other hand, the chains are cross-linked to a significant extent. When kerogen is heated to about 900° F (480° C), the links between the chains break and the solid is chemically transformed into an oil (representing about 60% of the kerogen weight), a fuel gas (9%), and a cokelike solid (25%). The resulting oil is heavy and stiff and has a high sulfur and nitrogen content. However, after further treatment it becomes as good a refinery feedstock as most crude petroleums. The composition of a typical high-grade oil shale is given in Table 5—7.

Oil-shale deposits are found in 30 states, but only those of the

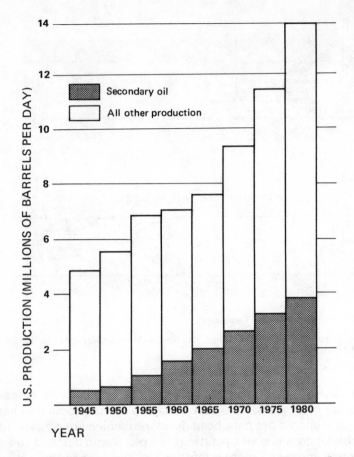

Figure 5—19.   Contribution of Secondary Recovery Techniques to Domestic Petroleum
Production

Redrawn from "Trends in Oil and Gas Exploration," Part 1, U.S. Senate Subcommittee
Hearings, Aug. 8—9, 1972, p. 335.

Green River Formation in Colorado, Utah, and Wyoming appear to be
commercially valuable at the present time. The deposits in this formation are located beneath 25,000 square miles (16 million acres) of land.
About 17,000 square miles of the deposit (11 million acres) are thought
to contain oil shale capable of development in the foreseeable future.
The known Green River Formation deposits, shown in Figure 5—20,
include high-grade shales representing approximately 600 billion barrels
of petroleum — an amount equal to about 200 times the current annual
domestic production. High-grade shale deposits are at least 10 feet thick
and average 25 or more gallons of petroleum per ton. An additional
200 billion barrels of petroleum are represented by lower grade shales
which are 10 feet thick and average 15—20 gallons per ton. The
recovery of even a fraction of the petroleum resources of the Green

Table 5—7.   COMPOSITION OF TYPICAL HIGH-GRADE OIL SHALE AVERAGING 25 GALLONS OF OIL PER TON

Component	Weight (percent)
Organic matter:	
Content of raw shale	13.8
Ultimate composition:	
Carbon	80.5
Hydrogen	10.3
Nitrogen	2.4
Sulfur	1.0
Oxygen	5.8
Total	100.0
Mineral matter:	
Content of raw shale	86.2
Estimated mineral constituents:	
Carbonates; principally dolomite	48
Feldspars	21
Quartz	13
Clays, principally illite	13
Analcite	4
Pyrite	1
Total	100

From "Final Environmental Statement for the Prototype Oil-Shale Leasing Program," vol. 1, U.S. Dept. of the Interior, 1973, p. 1-10.

River Formation would provide a significant supplement to the United States energy supply for many years.

The enormous size of the Green River deposits is readily seen when they are compared to known shale-oil reserves both inside and outside the United States. Such a comparison is given in Figure 5—21.

Two approaches to oil-shale development are under consideration: mining followed by surface treatment and in-situ (in-place) treatment. Only the mining-surface treatment approach has been developed to the point that commercial production is possible before 1980.

Oil shale may be mined by either surface or underground methods. When the overburden is not too thick, open-pit mining allows for nearly complete recovery of all exposed shale. When underground mining is used, room-and-pillar techniques enable a recovery of about 75% of the shale to be accomplished.

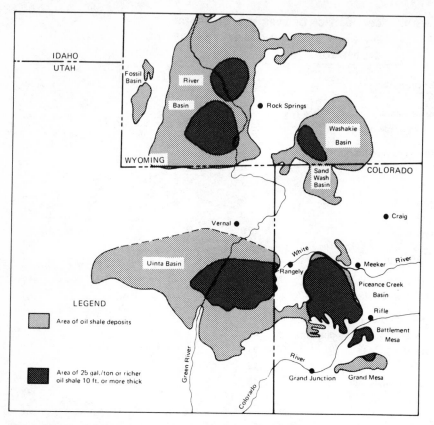

Figure 5–20.   The Green River Formation Oil-Shale Deposits

From "Final Environmental Statement for the Prototype Oil-Shale Leasing Program," vol. 1,
U.S. Dept. of the Interior, 1973, p. I-3.

After being mined, the shale is crushed and retorted at a temperature of 800°—1000° F (370°—540° C). Under these conditions, the solid
kerogen of the shale is converted into gas, oil vapors, and a char that
remains on the surface of the spent shale. A number of retorting
techniques have been patented but only three are considered to be good
candidates for early commercial use: (1) The gas combustion retort,
developed by the Bureau of Mines, (2) the union retort, developed by
the Union Oil Company of California, and (3) the TOSCO retort,
developed by the Oil Shale Corporation.

Any shale retorting technique presents problems that must be
overcome: a practical technique must involve the processing of many
tons of raw shale each day. The resulting large amount of stripped shale
presents disposal problems. As much of the vaporized oil as possible
must be condensed and recovered as a liquid. A large part of the heat
used in the process must be recovered and used again to keep fuel costs

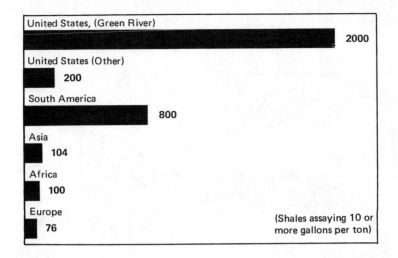

Figure 5—21. Known Shale Oil Resources of the World Land Areas, in billions of barrels

Data from "Organic-Rich Shale of the United States and World Land Areas" by Donald C. Duncan and Vernon E. Swanson, U.S. Geological Survey Circular 523, Washington D.C., 1965, pp. 9, 18.

minimized. A minimum amount of cooling water must be used since most shale is found in water-poor locations.

As an economy measure, the heat for the retorting process is obtained by burning the solid coke residue (char) left on the stripped shale or by burning a part of the generated gaseous products. One method for doing this is shown in Figure 5—22, which represents the TOSCO retort. This device is a rotary kiln in which heat is transferred to the shale by heated ceramic balls.

Water, an inherent by-product of oil-shale retorting, may be produced at a rate as high as 10 gallons per ton of shale. More typically, it ranges between 2 and 5 gallons per ton. This water together with the gas, oil, and coke produced accounts for 15%—20% of the weight of the original mined shale. Thus, the stripped shale (a disposal problem mentioned earlier) weighs 80%—85% as much as the original shale. Figure 5—23 contains a schematic diagram of a surface processing oil shale complex.

The Bureau of Mines and various commercial organizations have experimented with in-place processing that requires the underground heating of oil shale. The heating has been accomplished by underground combustion, by injecting heated gases, and by injecting superheated steam. The technology has not yet been developed to the point of predicting either technical or economic success. Two major problems have been encountered: (1) The natural permeability of the deposits is too low to allow for the passage of gases and liquids. Attempts to create

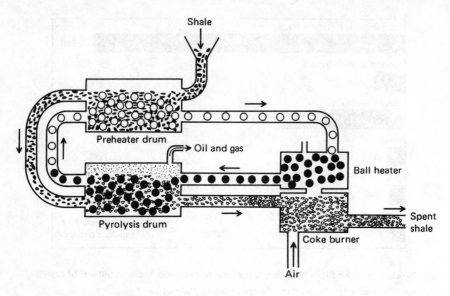

Figure 5—22.   The TOSCO Hot Ball Process

permeability artificially have met with little success. (2) The process cannot be controlled with sufficient accuracy through wells bored from the surface.

About 72% of the 11 million acres of commercially valuable oil shale deposits are located on public lands, managed by the federal government. Access to at least a portion of these deposits will be required before a significant oil-shale industry can develop. In July of 1971 the Department of the Interior made six tracts available for competitive bidding in a prototype oil-shale leasing program. This action followed a 41-year period in which no oil-shale lands were offered for lease. Bidding for the six tracts was delayed until late in 1973 in order to complete detailed environmental impact studies.

The Interior Department has projected shale oil production at no more than 100—300 thousand barrels per day (bpd) by 1983. Higher production levels require a greater access to public lands through revised leasing programs and significant technological advances, particularly in the area of in-situ processing.

The upper limit of shale oil production may be about one million bpd regardless of land availability and economic feasibility. This limit would be set by the supply of water. Three barrels of water are needed to produce one barrel of oil using existing technology. Surface water supplies in the Green River formation appear adequate for a one-million-bpd operation. An earlier estimated output of 3—5 million bpd would require all known surface water in these arid regions. Table 5—8

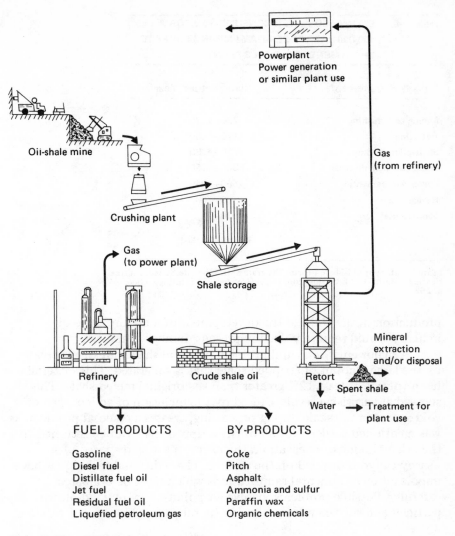

Figure 5—23. Oil Shale Surface Processing

Redrawn from "Final Environmental Statement for the Prototype Oil-Shale Leasing Program," vol. 1, U.S. Dept. of the Interior, 1973, p. I-8.

illustrates the water requirements for a 100-thousand-bpd plant. The development of in-situ technology would double the output ceiling since no water would be needed for spent shale disposal on the surface.

The availability of qualified miners and heavy mining equipment may also impose a limit on the development of the shale oil industry. Approximately 570 million tons of shale would be processed annually in a one-million-bpd plant. This is about equal to the annual coal

Table 5—8.   ESTIMATED WATER CONSUMPTION FOR A
100,000-BARREL-PER-DAY "MINING-SURFACE
TREATMENT" OIL-SHALE PLANT

Process Requirements	Acre-Feet per Year
Mining and crushing	730—1020
Retorting	1170—1460
Shale-oil upgrading	2920—4380
Processed shale disposal	5840—8750
Power requirements	1460—2040
Revegetation	0—700
Sanitary use	30—70
Total	12,150—18,420

From "Prototype Oil-Shale Leasing Program," U.S. Senate Subcommittee
Hearings, March 1974, Washington, D.C.: Government Printing Office,
p. 262.

production in the U.S., so the development of a mature shale oil
industry would require twice the present number of miners.

A major environmental problem of an oil shale industry is created
by the large amount of stripped shale that is generated. This material
has a volume at least 12% greater than the original mined shale. This
somewhat surprising result, caused by the inclusion of added open or
void spaces in the shale during processing, creates a disposal problem, as
was mentioned earlier. Not all of the stripped shale can be returned to
the original mine or excavation; the excess must be disposed of in
nearby canyons or piled on the surface. These disposal techniques have
important environmental ramifications which must be evaluated
carefully. Possible problems involve air pollution from wind-blown
particles and surface water pollution by salts leached from the residues.

Tar Sands

Tar sands, also known as oil sands and bituminous sands, constitute a potential source of petroleum. They differ from the more
conventional oil and gas reservoirs by virtue of the high-viscosity
hydrocarbons they contain. These hydrocarbons are generally solids or
semisolids at ordinary temperatures and therefore are not recoverable
by ordinary petroleum production methods. Reservoir energy is
typically minimal so that production must be initiated and sustained by
providing some sort of energy from outside the deposit.

The largest known tar sand deposit in the world is located at
Athabasca in Alberta, Canada. It covers an area of about 30,000 square
miles, and is estimated to contain material equivalent to approximately
174 billion barrels of synthetic crude petroleum. Technology has been

developed and is now in commercial use for obtaining petroleum from
the Athabasca deposits. The recovery process makes use of strip mining
and surface extraction. The first large-scale recovery project began in
1967 and produces 45,000 barrels of petroleum per day. A second large
project will reportedly begin in 1976. Further production increases
from this deposit are expected in the future as other projects are
approved and activated. By 1985 a production level of up to 1.25
million barrels per day could be reached.

In the present Canadian operation, tar sands are strip mined using
large excavators. The mixture of hydrocarbons, called *bitumen*, is
extracted from the sand by hot water. The bitumen is then heated in a
coking unit to produce coke, various liquid hydrocarbons and gas. The
coke is used for in-plant fuel and the liquid hydrocarbons are further
treated to produce a synthetic crude petroleum. A simplified diagram
of the process is given in Figure 5—24.

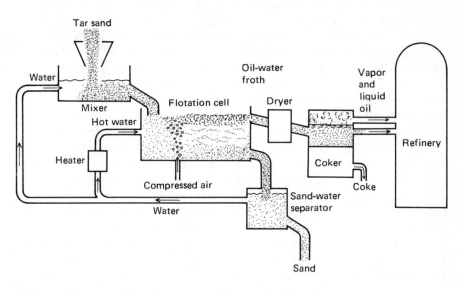

Figure 5—24. Extraction of Hydrocarbons from Tar Sands

From "Tar Sands and Oil Shales" by Noel de Nevers. Copyright Feb. 1966 by Scientific
American, Inc. All rights reserved.

The overall potential for obtaining petroleum from tar sands in the
United States is not well known. However, on the basis of the information available, it appears to be quite limited. The total estimated
recoverable petroleum in known deposits amounts to 19—29 billion
barrels. Most of this resource is found in five deposits, all located in the
state of Utah. At the present time these five deposits are the only ones
in the United States considered large enough to justify eventual
commercial development.

The use of Athabasca technology on the Utah deposits is very questionable because the Utah sands are believed to be much harder and therefore more difficult to process. The exploitation of Utah tar sands will probably require in-situ methods for which technology is now lacking. The lead time to develop the necessary technology and apply it to the deposits will probably prevent any significant petroleum production from Utah tar sands until after 1985. Furthermore, physical and other limitations to development make it apparent that total production from the Utah deposits is not likely to exceed 500,000 barrels per day.

Petroleum from Wastes

During each year in the United States, more than one billion tons of inorganic mineral wastes and about 2 billion tons of organic wastes are produced. At the present time a large percentage of this material is incinerated, dumped into waterways or used as landfill. These disposal methods are more and more often being attacked for contributing to environmental pollution. In addition, available space for landfills is running out. Research is now being conducted which could lead to a partial solution — the conversion of organic wastes into petroleum.

As early as 1971 proponents of waste conversion estimated that organic waste conversion would produce nearly 2.5 billion barrels of petroleum per year. This amounts to about one half the annual consumption in the United States. More recent estimates indicate the earlier figures were somewhat exaggerated. It is now known that more than one half the total weight of organic waste is water. In 1971 only 880 million tons of dry ash-free organic waste was produced, and 80% of it could not be collected for conversion. Manure or agricultural wastes, for example, cannot be collected economically from every small farm or ranch. It appears that the most readily collected organic wastes are those from large urban areas. The main sources of organic wastes in 1971 are given in Table 5—9 together with estimates of the amounts generated and the amounts considered to be readily collectable. As can be seen, only 136 million tons of a generated 880 million tons could be collected. This amount of waste would produce about 170 million barrels of petroleum — roughly 5% of the total amount consumed in 1973.

Three methods for converting wastes are being investigated seriously: hydrogenation, pyrolysis, and bioconversion. Bioconversion produces natural gas rather than petroleum products. Hydrogenation and pyrolysis have been advanced to the pilot plant and demonstration plant stages of development, respectively. It is likely that both processes will be commercialized before 1980. Bioconversion has not been subjected to as much research, and commercialization is not likely before 1985.

Hydrogenation, developed by U.S. Bureau of Mines scientists,

Table 5—9.　GENERATION OF DRY ORGANIC WASTES IN THE
UNITED STATES, 1971 (millions of tons)

Source	Wastes generated	Readily collectable
Manure	200	26.0
Urban refuse	129	71.0
Logging and wood manufacturing residues	55	5.0
Agricultural crops and food wastes	390	22.6
Industrial wastes	44	5.2
Municipal sewage solids	12	1.5
Miscellaneous	50	5.0
Total	880	136.3
Net oil potential (millions of barrels)	1098	170
Net methane potential (trillions of cubic feet)	8.8	1.36

Reprinted with permission from "Fuel from Wastes: A Minor Energy Source,"
by T. H. Maugh, II, *Science*, vol. 178 (Nov. 10, 1972): Table 1, p. 599.
Copyright 1972 by the American Association for the Advancement of Science.

involves as a principal reaction the extraction of oxygen from cellulose.
The wastes are reacted with carbon monoxide and steam at elevated
temperatures ($240°-380°C$) and pressures (100—250 atmospheres). An
85% conversion is normally obtained and the net yield is about 1.25
barrels of petroleum per ton of dry waste. This process might more
correctly be called *deoxygenation* or *chemical reduction* rather than
hydrogenation. The energy content of the resulting petroleum is about
15,000 Btu per pound; number 6 fuel oil from natural petroleum has an
energy content of about 18,000 Btu per pound.

The second major conversion method is pyrolysis or destructive
distillation. In this process organic waste is heated to about $500°C$ in an
oxygen-free environment. The waste breaks down under such conditions and an oil results — about one barrel per ton of waste. A major
disadvantage of the process is that in addition to oil, two other fuels —
gas and char — also result. This compounds the problems of collection
and marketing. However, construction and operating costs for the
pyrolysis process are anticipated to be much lower than for the
hydrogenation process because pyrolysis is performed at atmospheric
pressure.

Outer Continental Shelf
Development

It is estimated that the offshore areas of the United States contain
approximately 171 billion barrels of crude petroleum that is recoverable with existing technology. This represents approximately 40% of

the nation's total undiscovered petroleum (see Figure 5—2). The development of these offshore areas, commonly called the outer continental shelf (OCS), is a convenient way to increase domestic production and decrease the dependence on petroleum imports.

The federal government first made OCS land available for leasing in the early 1950s. The practice has been continued on a limited basis since then. Petroleum production from OCS sources now amounts to more than 400 million barrels annually. However, the use of offshore areas for petroleum production has been and continues to be a matter of controversy. Most objections are centered around problems of environmental protection.

The offshore areas around the United States are often classified according to the depth of water over them. The continental shelf includes those areas covered by water up to a depth of 200 meters (656 feet). Immediately beyond the continental shelf, under water that ranges in depth between 200 and 2500 meters (656—8200 feet), is the continental slope or outer continental shelf. Thus, the differentiation between the terms continental shelf and continental slope is based on water depth. It should be noted that the term outer continental shelf is often used loosely to denote all offshore areas.

Size estimates and the geographical distribution of the continental slope and shelf areas are given in Table 5—10. The total land area of the United States and its territories is 3.62 million square miles. The continental shelf is about 22% as large and the continental slope 13% as large as the total onshore area; they represent a substantial addition to the United States resource potential.

Estimates made in 1968 by the U.S. Geological Survey provide

Table 5—10.   OFFSHORE AREAS OF THE UNITED STATES
(thousands of square miles)

Location	Continental Shelf Area	Continental Slope Area
Hawaii	0.4	3.6
Alaska	560.0	212.2
Washington, Oregon and California coast	15.4	76.2
Gulf coast	107.5	84.2
Atlantic coast	122.0	102.5
Total	805.3	478.7

Data from "Outer Continental Shelf Policy Issues," Part 1 U.S. Senate Subcommittee Hearings, March 23—24, April 11, 18, 1972, p. 174.

information concerning the amounts, locations, and recoverability of offshore petroleum. These estimates are given in Table 5—11. Note that nearly equal total amounts of petroleum are estimated for the shelf and slope locations but that none of the slope resources are considered to be recoverable using known technology. The primary technological improvements needed to develop such deposits are in the category of deepwater operations.

Petroleum from the outer continental shelf has become an increasingly important part of domestic petroleum production. In 1973, more than 17.2% of the domestic production came from offshore sites compared to 4.5% in 1960. It is estimated that by 1985 as much as 25% of all domestic production will originate offshore. Steps were taken in 1973 to triple the amount of leased offshore acreage by 1979. Some of this acreage will be located in new frontier areas, including some under more than 200 meters of water.

Table 5—11.   ESTIMATED POTENTIAL OFFSHORE CRUDE PETROLEUM
RESOURCES OF THE UNITED STATES
(billions of barrels)

Location	Original Petroleum in Place	Amount Recoverable with Existing Technology
Alaska		
Continental shelf	217	54
Continental slope	156	0
Pacific		
Continental shelf	30	8
Continental slope	148	0
Gulf		
Continental shelf	240	60
Continental slope	189	0
Atlantic		
Continental shelf	169	42
Continental slope	143	0
Totals		
Continental shelf	660	160
Continental slope	640	0
Total	1300	160

Data from "Fuel Shortages," Part 1, U.S. Senate Subcommittee Hearings, Feb. 22, 1973, Washington, D.C.: Government Printing Office, p. 620.

The history of offshore petroleum production is indicated in Figure 5—25, where the continual increase in production is quite apparent. The leveling off and then decrease in production in 1972 and 1973 is directly related to environmental concerns. Most production to date has come from the waters off the Gulf of Mexico.

Opposition, based on environmental concerns, has become a major obstacle to the development of offshore petroleum resources. Since the 1969 blowout and petroleum spill in the Santa Barbara Channel and two accidents off the Louisiana shore in 1970, delays have been experienced by many production and exploration operations conducted off U.S. shores. It is the concern of some that environmental damage will result from many of the normal activities associated with offshore petroleum production, including exploratory surveying, platform placement, drilling, normal production and transportation, and storage. A discussion of the nature and magnitude of these environmental risks will conclude this chapter.

There is little evidence of any significant environmental damage resulting from geophysical explorations for petroleum. This is especially

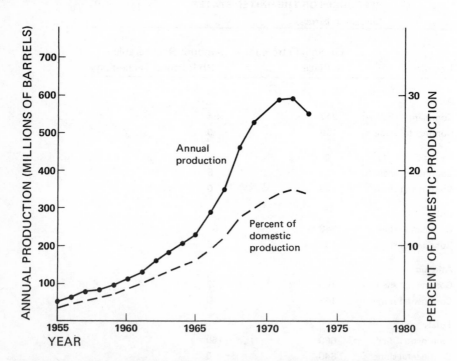

Figure 5—25.   Offshore Domestic Petroleum Production (includes state and federally owned lands)

Annual production data, 1955-69, from "Outer Continental Shelf Policy Issues," Part 1, U.S. Senate Subcommittee Hearings, March, April, 1972, p. 115. Percentages calculated from "Crude Petroleum — Domestic Reserves," *Chemical Economics Handbook*, p. 229.3500, Dec. 1972, Stanford Research Institute. Data for 1970-73 from *Oil and Gas Journal*, May 6, 1974, p. 136.

true since the use of explosives as an energy source has been largely discontinued in marine seismic operations. Less environmentally hazardous methods have been adopted for generating sound waves. These include the use of compressed air charges or vibrating acoustical systems. The only adverse effects from such surveys would be the resulting minor noise and exhaust emissions normally associated with diesel powered ships.

Water turbidity caused by the placement of drilling and production platforms involves only a small area and is of short duration. Destruction of the sea bottom is also confined to a small area — only a few square feet for each piling. Under certain conditions, structures can adversely affect commercial fishing activities. The increasing number of structures make traditional fishing grounds less suitable for fishing by trawlers. Also, platforms can pose hazards to commercial ships of all types that stray from established travel lanes. This hazard is minimized by the creation of safety fairways to permit the safe passage of ships in and out of port.

There is also concern about the disruption of scenic views by structures located within sight of coastal areas. This, it is thought, might affect tourist traffic. However, it has been pointed out that tourists might be attracted to the structures in search of the many species of fish which concentrate around them. The structures provide an artificial habitat for such fish and enhance sport fishing activities. There is some concern that normal predator-prey relationships might be upset by concentrating the fish this way.

Debris is composed of those substances that are discharged into the sea (excluding waste water) as the result of operations carried out on structures. The improper disposal of this type of material — trash, drilling mud, bilge waste, spilled crankcase oil, and engine fuel — can cause serious localized pollution problems. Toxic substances such as paints and thinners can poison and kill some organisms. Floating, nonbiodegradable debris is unsightly and inhibits the recreational uses of water. Fortunately, the amount of debris discharged into the environment from offshore operations is quite small because of the enforcement of appropriate regulations.

During the period from 1964 through 1972, thirty-nine oil spills involving 50 or more barrels occurred on federal offshore land. The total volume of petroleum involved has been estimated at slightly less than 300,000 barrels. During this same period, more than 2 billion barrels of petroleum were produced at offshore sites. Thus, the recorded spills represent about 0.014% of the total production during the period.

The greatest potential for blowouts exists during exploratory drilling activities. In addition to water pollution, air is sometimes polluted by products from the fires that often accompany such misfortunes. Normal procedures usually protect against blowouts, but they do sometimes occur. During the sixteen years from 1956 to 1972, forty-one blowouts took place on federal offshore land. Ten of these resulted in water-polluting petroleum spills.

The reports of several major studies vary in their assessment of the damage that results from marine-located petroleum spills. Some reports conclude that no permanent damage results, while others insist that great harm has been done both immediately and over the long term. Birds living in the open sea are frequently the most obvious victims because they are likely to come into direct contact with the spilled petroleum. Oil contamination destroys the waterproof quality of their feathers, a condition from which they seldom recover even when careful rehabilitation is attempted. Under proper conditions of wind and current directions, nearshore and coastal environments can be seriously affected by spills that occur in offshore locations. Beaches, water recreation areas, and historic sites can be rendered temporarily unuseable. A more critical problem, however, is the degradation of estuarine and marsh areas which are vital to the ecology as breeding and nursery grounds.

Waste water from offshore petroleum production can be a serious source of water pollution. Water, present in a petroleum reservoir formation, is often recovered along with the petroleum. This so-called waster water is separated from the petroleum and discarded. Unfortunately, the separation of petroleum and water is not complete and some petroleum remains in the discarded water. The petroleum content of waster water discharged during offshore operations is regulated at an average of not more than 50 parts per million (.005%). Little research has been done to determine the effects of this level of pollution on the environment.

Offshore pipelines are normally buried — those located in 200 feet or less of water are required by law to be buried. This is done to avoid damage from dragging ship anchors or from movement caused by strong water disturbances resulting from intense storms. Approximately 98% of offshore petroleum is transported to shore by pipelines. The remaining 2% is transported by barge.

Well blowouts attract the most attention, but petroleum spills from ruptured pipelines can be just as serious. Broken pipelines were responsible for more pollution during the 1963—73 decade than all offshore drilling and production operations combined. The largest spill, 160,639 barrels, was caused by a pipeline leak resulting from anchor dragging by a ship. This accident took place on October 15, 1967.

In summary, it seems that even with the best systems and controls some pollution from offshore petroleum production will occur. Recently strengthened regulations and operating orders are as stringent as present technology allows. Natural disasters, equipment failure, or human error can take place despite regulations and enforcement procedures designed to minimize environmental damage from offshore operations.

Suggested Readings

Berg, R. R., J. C. Calhoun, Jr., and R. L. Whiting, "Prognosis for Expanded U.S. Production of Crude Oil," *Science*, April 19, 1974, pp. 331–336.

Bureau of Mines Information Circular 8549, "Energy Potential from Organic Wastes: A Review of the Quantities and Sources," Government Printing Office, Washington, D.C.

De Nevers, N., "The Secondary Recovery of Petroleum," *Scientific American*, July 1965, pp. 35–42.

————, "Tar Sands and Oil Shales," *Scientific American*, Feb. 1966.

Department of the Interior, "Final Environmental Impact Statement of Proposed Trans-Alaska Pipeline," Government Printing Office, Washington, D.C., 1972.

————, "Final Environmental Statement for the Prototype Oil Shale Leasing Program," Government Printing Office, Washington, D.C., 1973.

————, "Plain Facts About Oil," Government Printing Office, Washington, D.C., 1963.

Frank, H. J., "Economic Strategy for Import-Export Controls on Energy Materials," *Science*, April 19, 1974, pp. 316–321.

Metz, W. D., "Oil Shale: A Huge Resource of Low-Grade Fuel," *Science*, June 21, 1974, pp. 1271–1275.

"Oil and Gas Imports Issues," Hearings before the U.S. Senate Committee on Interior and Insular Affairs, January 1973, Government Printing Office, Washington, D.C.

"Outer Continental Shelf Policy Issues," Hearings before the U.S. Senate Committee on Interior and Insular Affairs, March/April 1972, Government Printing Office, Washington, D.C.

Rose, S., "The Far-Reaching Consequences of High-Priced Oil," *Fortune*, May 1974, p. 106.

Stauffer, T. R., "Oil Money and World Money: Conflict or Confluence," *Science*, April 19, 1974, pp. 321–325.

"Trends in Oil and Gas Exploration," Hearings before the U.S. Senate Committee on Interior and Insular Affairs, August 1972, Government Printing Office, Washington, D.C.

Chapter 6
Natural Gas

The first significant use of gaseous fuel took place in London,
England, near the beginning of the nineteenth century. Manufactured
gas obtained by heating coal was used in this pioneering event. The
initial use was limited to industrial applications, but gas street lighting
soon followed. The first gas distribution network resulted from a street
lighting franchise granted in 1812 to the Westminster Gas Light and
Coke Company of London. The popularity of manufactured gas for
lighting spread rapidly, and in 1816 the first United States franchise
was granted in Baltimore, Maryland. By 1859 a gas company existed
in almost every city of importance.

The first recorded attempt to develop natural gas supplies in the
United States occurred in 1821. A pitlike well was dug to utilize a source
of natural gas seepage located near Fredonia, New York. The seepage
had been discovered through accidental ignition that demonstrated the
excellent fuel characteristics of the gas. A crude system of hollowed
logs was used to contain the gas in the pit. It was piped from the pit to
a nearby inn where it was used for cooking, heating, and lighting.
From this effort evolved the first natural gas corporation in America,
the Fredonia Gas Light Company founded in 1858. The desirable
characteristics of natural gas and an increase in the number of local
sources of supply created an increased demand. However, the use of
natural gas in areas located long distances away from the sources was
restricted by a lack of transportation methods.

The inability to transport natural gas caused a continued increase
in the demand for manufactured gas. This increase was further enhanced
by improvements in the heat content of manufactured gas and the
technology involved in handling gaseous fuels. The quantity of
manufactured gas sold increased from 102 billion cubic feet in 1901
to 421 billion cubic feet in 1925. During this same period of time
the number of customers increased from 4.2 million to 10.6 million.
Sixty-five percent of the gaseous fuel customers of 1932 had service
that involved manufactured gas. This was in spite of the fact that
natural gas comprised 80% of the gas sales at that time. Most of these
sales, however, were to large industrial users located near the sources

of natural gas. During the 1940s, the economics involved began to change. The cost of manufactured gas raw materials rose sharply and substantial quantities of natural gas were discovered. In addition, gas transporting technology improved markedly — large-diameter, high-pressure pipelines were developed. The use of natural gas in all parts of the country became economically feasible. Manufactured gas use peaked in the mid-1940s. The highest number of customers, 11 million, was served in 1946 and the largest volume of gas, 760 billion cubic feet, was sold in 1947. In 1967 less than 670,000 customers were involved with manufactured gas service. Today, manufactured gas sales represent a negligible amount of total gas sales — less than 1%.

Natural gas use has grown faster than that for any other primary fuel in the United States since World War II. During the period 1950—70 the consumption of natural gas increased at an annual average rate of 6.6%. During the same time, the rates of increase in petroleum and coal consumption were respectively 3.9% and 0.3%. About one third of all energy now used in the United States comes from the combustion of natural gas. Representations of the increasing role of natural gas in U.S. energy consumption are given in Figures 1—3 and 3—3.

Origin and Occurrence of Natural Gas

Natural gas is thought to have originated from materials and processes similar to those responsible for petroleum formation — the decomposition of marine organisms (both plant and animal) in the absence of oxygen in the environment of deep ocean-bottom sediments. Natural gas occurs in reservoirs under geological conditions similar to those associated with petroleum deposits: (1) adequate reservoir rock of appropriate porosity and permeability, (2) a relatively impervious cap rock, and (3) a structural trap to confine the natural gas. The structural traps are of the same general types as those for petroleum (see Figure 5—1). Because of these similarities in reservoir conditions, natural gas and petroleum are often, though not always, found together in the same reservoir. When they are found together, the natural gas is referred to as *associated natural gas*, while natural gas found alone is called *nonassociated natural gas*.

Chemical Composition of Natural Gas

The simple hydrocarbon methane, CH_4, is the predominant component of commercial natural gas supplied to consumers. It is found in amounts representing 80%—95% of the natural gas consumption. Some ethane, C_2H_6, is almost always present together with smaller amounts of other heavier saturated hydrocarbons — propane, butanes, pentanes, etc. Certain nonhydrocarbon gases such as nitrogen and carbon dioxide are also common components. The concentration of these minor components in natural gas varies slightly with the source of the gas and, as shown in Table 6—1, affects the heating value.

Natural gas as obtained from underground deposits generally has a composition that is significantly different than that of the familiar commercial fuel. The crude gas usually contains some undesirable impurities and, in addition, some heavy, condensable hydrocarbons. Appropriate processing eliminates or reduces the amount of the undesirable impurities and allows the condensable hydrocarbons to be collected as a separate fraction of industrial value.

Crude natural gas is usually classified into one of two broad categories on the basis of chemical composition. *Wet natural gas* contains significant amounts of condensable hydrocarbons while *dry natural gas* contains amounts that are too small for economical recovery. Wet gas is converted to dry gas by processing which removes the condensable hydrocarbons. The recovered condensable hydrocarbons are called natural gas liquids (NGL); the propane and butanes obtained from these liquids are marketed as *liquefied petroleum gas* (LPG or LP gas). Pentanes and heavier hydrocarbons obtained from natural gas liquids are referred to as *natural gasoline*. This material evaporates readily and is usually blended with gasoline obtained from petroleum refineries to produce a more volatile gasoline, which ignites readily in cold weather.

Crude natural gas may also be classified as sour or sweet on the basis of its composition. *Sour natural gas* contains sufficient amounts of sulfur compounds and/or carbon dioxide to be corrosive or offensive smelling. *Sweet natural gas* does not have these properties and can be obtained from sour gas by processes which remove the offending components. Table 6—2 contains data which illustrate the

Table 6—1.   COMMERCIAL NATURAL GAS COMPOSITION AND HEATING VALUE

| Source | Composition (volume percent) | | | | | | | Heating Value (Btu/ft^3) |
	Methane (CH$_4$)	Ethane (C$_2$H$_6$)	Propane (C$_3$H$_8$)	Butanes (C$_4$H$_{10}$)	Pentanes (C$_5$H$_{12}$)	CO$_2$	N$_2$	
Birmingham, Alabama	93.14	2.50	0.67	0.32	0.12	1.60	2.14	1024
Columbus, Ohio	93.54	3.58	0.66	0.22	0.06	0.85	1.11	1028
Dallas, Texas	86.30	7.25	2.78	0.48	0.07	0.63	2.47	1093
Denver, Colorado	81.11	6.01	2.10	0.57	0.17	0.42	9.19	1011
New Orleans, Louisiana	93.75	3.16	1.36	0.65	0.66	0.42	0.00	1072
San Francisco, California	88.69	7.01	1.93	0.28	0.03	0.62	1.43	1086

Reprinted with permission from *Gas Engineers' Handbook* (New York: Industrial Press, Inc., 1965), Table 2—9 on p. 2/10.

varied compositions and classifications of natural gas from different sources.

The undesirable impurities found in crude natural gas include carbon dioxide, nitrogen, water vapor, hydrogen sulfides, and thiols, or other organic sulfur compounds. Carbon dioxide and nitrogen usually reduce the heating value of the gas — in some cases to the extent that the gas cannot be marketed until they are reduced or removed. Some natural gases contain enough carbon dioxide to allow for economical recovery and conversion into dry ice. Water vapor impurities in natural gas create some unique problems. If it is present in high enough concentrations, it can condense as a result of temperature

Table 6–2.   CRUDE NATURAL GAS COMPOSITIONS AND CLASSIFICATIONS

Composition (volume percent)

	Rio Arriba County, N.M.	Terrell County, Texas	Stanton County, Kansas	San Juan County, N.M.	Olds Field, Alberta, Canada	Lacq Field, France	Cliffside Field, Amarillo, Texas
Methane	96.91	45.64	67.56	77.28	52.34	70.00	65.8
Ethane	1.33	0.21	6.23	11.18	0.41	3.0	3.8
Propane	0.19		3.18	5.83	0.14	1.4	1.7
Butanes	0.05		1.42	2.34	0.16	0.6	0.8
Pentanes and heavier	0.02		0.40	1.18	0.41		0.5
Carbon dioxide	0.82	53.93	0.07	0.80	8.22	10.0	
Hydrogen sulfide		0.01			35.79	15.0	
Nitrogen	0.68	0.21	21.14	1.39	2.53		25.6
Helium							1.8
Total	100.00	100.00	100.00	100.00	100.00	100.00	100.0
Total sulfur, grains/100 ft³	0	6.3	0	0	22,525		
Classification							
Wet				x			
Dry	x	x	x		x	x	x
Sweet	x		x	x			x
Sour		x			x	x	
Gross heating value, Btu/ft³	1010	466	938	1258	807	921	825

Reprinted with permission from *Encyclopedia of Chemical Technology*, 2nd ed., by R. E. Kirk and D. F. Othmer (New York: John Wiley and Sons, Inc., 1966), vol. 10, p. 450.

drops or pressure increases. The resulting liquid decreases the efficiency of the pipeline containing it. Under certain conditions of temperature and pressure, water forms hydrates with the hydrocarbons present in natural gas. These hydrates are similar in appearance to ice crystals, and if allowed to form, they can greatly restrict or actually stop the flow of gas through a pipeline. Sulfur compounds, mainly H_2S and thiols, are objectionable impurities because they create both corrosion and odor problems — they are acidic substances with pungent odors. If a sufficient quantity of hydrogen sulfide exists in crude natural gas, it can be removed and used as a source for elemental sulfur, which is a valuable chemical. Helium is found in a few natural gases, but it is considered to be a valuable component rather than an impurity when it is present at levels high enough to allow economical recovery (about 0.3%—0.7% or higher).

Natural Gas Processing and Purification

The removal of heavy hydrocarbons from wet natural gas is often accomplished by scrubbing the gas with a heavy hydrocarbon oil. The heavier hydrocarbon vapors, being more soluble in the scrubbing oil than methane and ethane, dissolve and are removed. The typical equipment involved in an oil-absorption process consists of a tower for absorbing the propanes, butanes, and other heavier hydrocarbons, a stripping tower for removing these hydrocarbons from the scrubbing oil, and a purification system for separating the recovered hydrocarbons which are then converted into marketable materials (propane, butane, natural gasoline, etc.).

Water is removed from natural gas by adsorption onto the surface of activated solid drying agents (desiccants) or by absorption into hygroscopic (moisture-retaining) liquids. The use of hygroscopic liquids in an absorption system is generally the most economical method when large volumes of natural gas are involved. Many liquids possess dehydrating properties but only two, diethylene glycol and triethylene glycol, are used extensively in the natural gas industry. The structures of these two compounds are given in Table 6—3.

A glycol dehydration unit consists primarily of an absorption tower in which the glycol stream flows countercurrent to the gas stream, a regenerator where the absorbed water is distilled from the glycol, and a solvent reclaimer which is used to prevent the buildup of undesirable impurities in the glycol.

Carbon dioxide and hydrogen sulfide may be removed from natural gas in a variety of different ways. The method most often used in the United States is the Girbotol process. In this process aqueous solutions of monoethanolamine (and under some circumstances diethanolamine) are used. These two ethanolamines, whose

Table 6—3.   COMPOUNDS USED TO PROCESS AND PURIFY NATURAL GAS

Dehydrating Agents

$HO-(CH_2)_2-O-(CH_2)_2-OH$	$HO-(CH_2)_2-O-(CH_2)_2-O-(CH_2)_2-OH$
diethylene glycol	triethylene glycol

CO_2 and H_2S Removal Agents

$HO-(CH_2)_2-NH_2$	$HO-(CH_2)_2-NH-(CH_2)_2-OH$
monoethanolamine	diethanolamine

Odorizing Agents

C_2H_5-SH	C_4H_9-SH
ethanethiol	a-butanethiol

structures are given in Table 6—3, differ in their selectivity for H_2S and CO_2. It is this selectivity that determines the combination used. The general reaction of H_2S with these substances is reversible upon heating and can be represented as:

$$2RNH_2 + H_2S \longrightarrow (RNH_3)_2S \qquad \text{6-1}$$

The apparatus used in the process consists of an absorption tower, a stripping tower, and various pumps, condensers, and heat exchangers. A combination of the glycol process and the ethanolamine process, known as the *glycol-amine process*, is widely used in the natural gas industry. Mixtures of di- or triethylene glycol and monoethanolamine containing about 5% water are used as solvents for simultaneously dehydrating natural gas and removing acid gas constituents such as H_2S and CO_2.

No economical process has been developed to remove nitrogen from most natural gases. In those few instances where helium is removed by cryogenic (very low temperature) means, nitrogen can be removed as well.

With few exceptions, processed natural gas is not toxic to plant or animal life. For example, animals kept in an atmosphere containing 25% natural gas for 30 days remained normal in every respect. An 8-hour exposure to an 80% natural gas atmosphere left another group of animals unaffected. In most cases plants and tree roots exposed to natural gas were not injured.

Odorizing compounds such as the thiols of ethane, butane, and pentane or other organic sulfides or disulfides are added to natural gas in small amounts to allow gas leaks to be detected by smell. This is important, since natural gas, while not toxic, forms dangerous explosive mixtures with air. Table 6—3 gives the structures of representative odorizing compounds.

Natural Gas Resources and Reserves

An overview of the domestic natural gas situation can be obtained by considering five quantities:

1. The total potential natural gas in place (the total of the volume of natural gas that has been discovered and the estimated volume that is still to be discovered).
2. The future potential natural gas in place (the estimated volume of natural gas that is still to be discovered).
3. Original natural gas reserves in place (the total estimated volume of natural gas contained in presently known reservoirs prior to any production, including productive reservoirs, depleted reservoirs, and known reservoirs from which there has been no production to date).
4. Cumulative production of natural gas (the sum of the estimated production for the current year and the actual production for each of the prior years).
5. Proven reserves of natural gas (the amount of natural gas estimated to be recoverable in the future from known reserves under existing economic and operating conditions).

The magnitude of each of these quantities is given in Figure 6–1, where the units used are trillions of cubic feet. The production unit in the natural gas industry is a volume (cubic foot), since the volume of a gas is more conveniently determined than the weight. Because gas volumes change with temperature and pressure, these two parameters must be specified in order for a volume measurement to have meaning. The volumes used in the natural gas industry are measured under a pressure of one atmosphere (14.7 lbs/in.^{2}) and at a temperature of 60°F (15.6°C).

The quantities shown in Figure 6–1 are defined such that mathematical relationships exist between them. For example, A = B + C and C = D + E. An estimate of the maximum amount of natural gas available in the future is equal to the sum of B and E, the future potential plus the proven reserves. This amounts to 1396.1 trillion cubic feet which is 76% of the total potential natural gas in place. Cumulative production accounts for the other 24% of total potential gas in place.

Estimates of future potential in-place natural gas are based mostly on geological data which take into account the volume of sediments judged to be favorable for the accumulation of natural gas. This potential represents reserves that are expected to be found by test wells drilled in the future under assumed conditions of adequate but reasonable prices and normal improvements in technology. Such normal technological improvements do not include major break-throughs. For example, the estimated 320 trillion cubic feet of natural

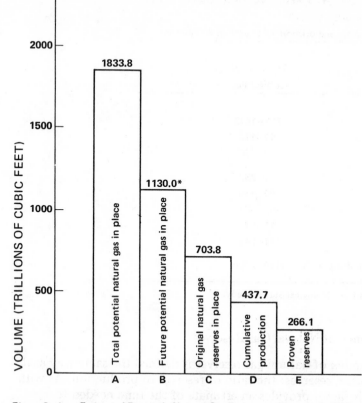

Figure 6—1. Estimated Domestic Natural Gas Resources, January 1, 1973

*Future potential natural gas in place is estimated at 1130—2250 trillion ft³. We have used the lower extreme of the range. — Eds.

Data from "Trends in Oil and Gas Exploration," Part 1, U.S. Senate Subcommittee Hearings, Aug. 8—9, 1972, pp. 428, 429, and U.S. Geological Survey study.

gas located in low-productivity reservoirs of the Rocky Mountain area are not included in the future potential estimate. Nuclear fracturing, a technique discussed later in this chapter, represents a possible method for improving the production of these reservoirs, but it is not considered to be a normal technological improvement. It will be a major breakthrough — if and when it is proven to be effective and safe. A further analysis of potential future in-place natural gas is given in Table 6—4 in terms of offshore and onshore locations.

We see from these data that a substantial portion of the undiscovered potential gas lies in offshore areas (37%) or in Alaska (10%). Such locations will be more difficult and expensive to explore and develop, and will also impose increased transportation costs on gas deliveries to consumers. Further details related to the proven reserve part of the natural gas resource base are given in the next two sections.

Table 6—4. FUTURE POTENTIAL NATURAL GAS IN PLACE,
JANUARY 1, 1973
(onshore and offshore to a depth of 200 meters)

	Trillions of Cubic Feet
Onshore	**712—1415**
Conterminous states	593—1177
Alaska	119—238
Offshore	**418—835**
Atlantic	55—110
Gulf of Mexico	181—361
Pacific	11—22
Alaska	171—342
Total onshore and offshore	**1130—2250**

Data selected from U.S. Geological Survey study.

Significance of Natural Gas Reserves

The life index of natural gas reserves is defined the same way as it was for petroleum reserves: the ratio of reserves to production. As with petroleum, this factor provides an estimate of the time needed to exhaust the reserves, assuming no new reserves are discovered and production levels remain unchanged. Life indexes of 50.0 and 11.1 years are obtained for potential and proven reserves respectively when the data of Figure 6—1 are used together with a 1973 production figure of 22.6 trillion cubic feet.

The steady decline of the natural gas life index has generated concern. The extent of this decline can be seen in Figure 6—2. Immediately after World War II, the life index exceeded 30 years. At that time, the market for natural gas was limited by the lack of an interstate pipeline delivery system, and substantial quantities of new natural gas were discovered and added to the reserve inventory as a result of petroleum exploration. However, natural gas production accelerated rapidly after the postwar construction of pipeline distribution systems, and the life index began to decline. It dropped below 20 years for the first time in 1961 and is now approaching 10 years.

The life index is influenced by both the annual production rate of natural gas and the increase or decrease in proven reserves that occurs during any year. It increases or decreases depending on whether production is less than or greater than the amount of additional

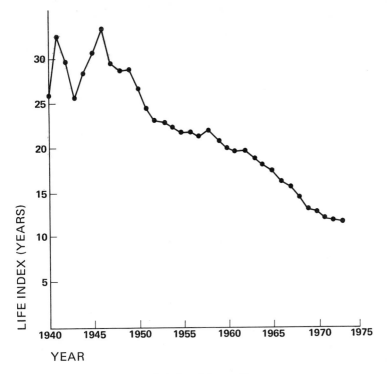

Figure 6—2.   Natural Gas Life Index of Proven Reserves

Calculated from "Natural Gas Reserves," p. 229.2011, Dec. 1972, in *Chemical Economics Handbook*, Stanford Research Institute. 1972-73 data calculated from Fig. 6-3.

reserves proven during the year. Figure 6—3 shows how these factors have varied historically. These data show that the proven reserves steadily increased (nearly tripled) from 1940 until 1967. They then decreased in 1968 and 1969 but increased again in 1970 because of the large reserves discovered in Alaska. Since 1970 they have decreased.

Net production increased steadily through 1970 and then leveled. Production in 1970 was more than 6 times greater than that of 1940 (an increase of 600%). The amount of new reserves added each year fluctuated a great deal but always exceeded the production until 1968. Since then, they have dropped below production except during 1970 when the Alaskan reserves were added. The net effect of this trend (production increasing faster than new reserves) is the previously noted decrease in life index.

Location of Proven Reserves

The location of proven natural gas reserves will be described from the geographical and geological points of view. The geographical

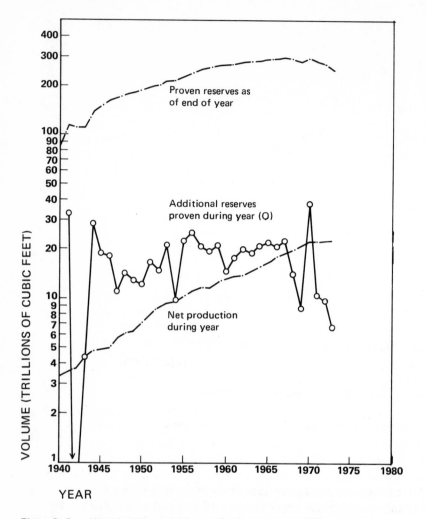

Figure 6–3.   Historical Trends in Natural Gas Production and Discovery

Calculated from "Natural Gas – Reserves," p. 229.2010, Dec. 1972 in *Chemical Economics Handbook,* Stanford Research Institute. Data for 1973 are from *Oil and Gas Journal,* April 1, 1974, p. 43.

approach, of course, involves the distribution of reserves among the various states. The geological approach relates to the conditions under which the reserves are stored.

Thirty-four states contain proven natural gas reserves. However, most reserves are found in just a few states. Almost 90% of the reserve total is located in 6 states and more than 60% is found in 2 states, Texas and Louisiana. These facts are shown by the data of Table 6–5.

From a geological point of view, we can look at the reserves in terms of the types of formations in which they are found. On this

Table 6–5.   ESTIMATED TOTAL DOMESTIC PROVEN
NATURAL GAS RESERVES, JANUARY 1, 1974

Location	Total Reserve (trillions of cubic feet)	Percent of Total
Texas*	84.9	34.0
Louisiana*	69.2	27.7
Alaska	31.6	12.7
Oklahoma	14.1	5.6
New Mexico	12.5	5.0
Kansas	11.7	4.7
California*	5.2	2.1
Wyoming	4.1	1.6
Other states	16.7	6.7
Total	250.0	100.0

*Includes offshore reserves.

Data from *Oil and Gas Journal*, vol. 72, no. 13 (April 1, 1974): 42.

basis reserves will be classified into one of three categories: non-associated, associated, and underground stored.

We remember that nonassociated gas is found in wells that contain no petroleum while associated gas is found along with petroleum. Of course, both of these classifications involve underground storage but our third category, underground stored, refers to gas that has been placed into underground storage after production. This practice is used to make natural gas available during peak demand periods. For example, more gas is needed in the winter than in the summer. A pipeline for winter use only is impractical, so gas is brought through pipelines during periods of low demand and stored near the consuming area for use during times of high demand.

Several storage techniques are used. In one called *line pack storage*, gas is stored under high pressure in the pipeline itself; the pipeline becomes an enormous reservoir. The most striking example of underground storage involves the use of exhausted gas or oil fields as underground reservoirs. Natural gas is pumped into an exhausted field located near the consumers and withdrawn as needed. Sometimes, porous rock formations are used for underground gas storage even if they never before held gas.

Regardless of the type of underground reservoir used, it must be properly prepared prior to use. The field must be cleaned by flushing with an inert gas and reconditioned. All abandoned boreholes must be located and plugged.

Most underground storage facilities are located in the northeastern United States. Pennsylvania has 66 dry-gas fields converted to storage, West Virginia has 31, Michigan has 28, and 15 each are located in New York, Kansas, and Kentucky.

A total of 1.8% (4.1 trillion cubic feet) of all proven natural gas reserves were located in underground storage as of January 1, 1974. The majority of proven reserves — 68.9% or 172.3 trillion cubic feet — is classified as nonassociated, while 29.4% (73.6 trillion cubic feet) is associated gas.

Production and Consumption of Natural Gas

The following important generalizations can be made concerning domestic natural gas production:

1. The majority of natural gas comes from gas wells (nonassociated gas) rather than oil wells (associated gas).
2. The majority of natural gas comes from onshore locations.
3. The majority of natural gas comes from a very small geographical area of the United States.

Figure 6—4 shows the distribution of gas production between types of wells (oil versus gas) and location of wells (onshore versus offshore). It is apparent that onshore gas wells form the backbone of the natural gas industry.

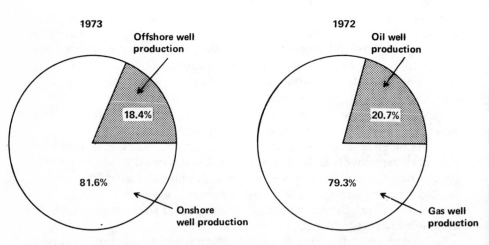

Figure 6—4.   Distribution of Natural Gas Production

Data from *Minerals Yearbook — 1971*, vol. 1, U.S. Dept. of the Interior, Bureau of Mines, p. 783 and from "Relationship of Energy and Fuel Shortages to the Nation's Internal Development," U.S. House of Representatives Subcommittee Hearings, Aug. 1972, p. 381.

Natural gas production is concentrated in a very small geographical area. This is a natural result of the previously discussed skewed distribution of proven reserves (see Table 6—5). Table 6—6 illustrates the extent of the concentration of production in a few states. Texas and Louisiana, which contain 61.7% of domestic proven reserves, account for an even greater share of domestic production — 73.8%. In 1973 Louisiana moved ahead of Texas as the leading state in natural gas production. Oklahoma and New Mexico are the only other states which account for more than 5% of total domestic production of natural gas. A comparison of the producing states and those containing reserves (Tables 6—6 and 6—5) reveals an identical rank ordering with the exception of Texas, Louisiana, and Alaska. Even though it is ranked third in size of reserves, Alaska is not among the leading producing states. Production from Alaskan reserves is anticipated to begin in the near future. It should be recalled that a similar situation exists relative to Alaskan petroleum reserves and production.

The limited geographical distribution of production does not carry over into consumption. A strong nationwide demand exists for natural gas, and it has resulted in the construction of a vast pipeline distribution system which is represented in Figure 6—5. Details related to the distribution and consumption of natural gas are given in Figures 6—6 through 6—8. In Figure 6—6, a comparison is made between natural gas production and consumption for nine regions. The states included in each region are shown in Figure 6—7. Figure 6—8 shows the movement of gas between regions.

Table 6—6.   DOMESTIC NATURAL GAS
PRODUCTION, 1973
(trillions of cubic feet)

State	Production	Percent of Total Production
Louisiana	8.46	37.4
Texas	8.24	36.4
Oklahoma	1.78	7.9
New Mexico	1.19	5.3
Kansas	.90	4.0
California	.48	2.1
Wyoming	.38	1.7
Other states	1.18	5.2
Total	22.61	100.0

From *Oil and Gas Journal,* April 1, 1974, p. 43.

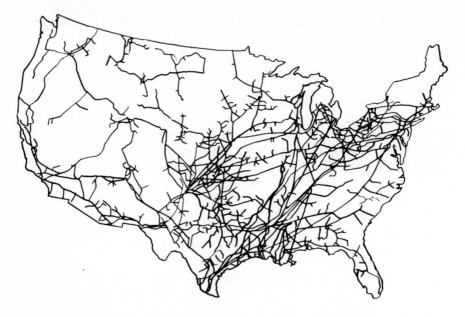

Figure 6—5.   Major Natural Gas Pipelines in the Conterminous United States

Redrawn with permission from *The Story of Gas* by Elbert C. Weaver, p. 11, copyright American Gas Association, Inc.

The dominant role played by the west south central region is apparent in both natural gas production and consumption. Production is at least 10 times that of any other region, and consumption is nearly double that of the next largest consumer region. The mountain region is the only one other than the west south central that produces more gas than it consumes, and the excess production is small. Thus, one region is essentially the natural gas source for all domestic consumption. In Figure 6—8 the width of each arrow is proportional to the quantity of gas being moved. A small amount of importing and exporting is involved. The subject of imports is discussed later in this chapter.

Nearly all of the marketed natural gas (98.1%) is used as a fuel or to generate electrical power. The small remainder (1.9%) is used as a chemical raw material in the petrochemical industry. Table 6—7 contains a summary of natural gas use in these categories.

We see that industrial uses account for a little more than two thirds of the consumption for fuel and power generation. This includes natural gas used in electrical power generating plants (3993 billion ft^3), petroleum and natural gas industry field use (2297 billion ft^3), petroleum refining (1063 billion ft^3), and pipeline fuel for moving the gas

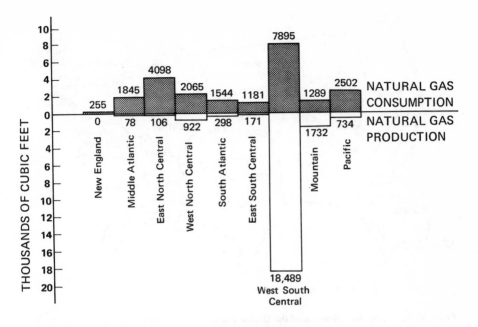

Figure 6—6. Regional Natural Gas Production and Consumption, 1971

Data from *Minerals Yearbook — 1971*, vol. 1, U.S. Dept. of the Interior, Bureau of Mines, pp. 784, 785, 792, and 793.

itself (743 billion ft^3). Residential uses are those normally associated with our residences: comfort heating and cooling, cooking, water heating, etc. Commercial uses include those similar to residential use, as well as others unique to establishments engaged in wholesale or retail trade.

The data of Table 6—7 imply that natural gas is not very important as a raw material for the petrochemical industry. This is true if we limit ourselves to natural gas that has been prepared for consumption.

Table 6—7. NATURAL GAS CONSUMPTION ACCORDING TO END USE, 1971 (billions of cubic feet)

Fuel and Power	22,245
Industrial	15,100
Residential	4,972
Commercial	2,173
Raw Material	432

Data selected from "Petroleum and Natural Gas — 1971 Flowchart," p. 229.1000C, Jan. 1974 and "Natural Gas — Consumption," p. 229.2030B, March 1973, in *Chemical Economics Handbook*, Stanford Research Institute.

Note: Pacific region includes Hawaii.

Figure 6–7.   Regional Divisions of the United States

From *United States Energy Fact Sheets, 1971*, U.S. Dept. of the Interior, Feb. 1973, p. 128.

This material is composed primarily of methane which is not used to a great extent as petrochemical feedstock. The important contribution of natural gas to the petrochemical industry is in the form of condensable hydrocarbons removed during the processing and purification steps. These compounds (ethane, propane, and butanes) are extremely important starting materials used in the preparation of many petrochemical products. These products, as mentioned in Chapter 5, are very important in our modern society and include such things as plastics, medicines, and synthetic textiles.

Problems Related to Natural Gas Use

The growth rate of natural gas demand is one of the main problems related to its use. This growth rate was acceptable and presented few problems until the decade of the 1970s. Now, however, supply-demand imbalances are becoming increasingly evident. Problems are developing which are similar to those discussed for petroleum in Chapter 5. Some of these problems are (1) a rapidly decreasing life index, (2) an increasing interest in imports, and (3) decreased drilling activity aimed at finding new domestic sources.

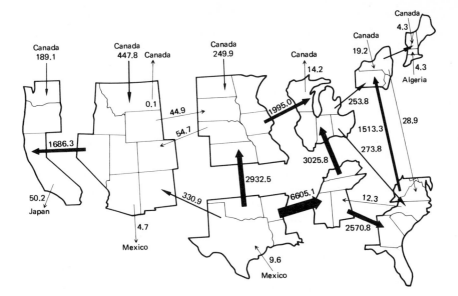

Figure 6—8.   Natural Gas Movement Between Regions (billions of cubic feet)

Data from *Minerals Yearbook — 1971,* vol. 1, U.S. Dept. of the Interior, Bureau of Mines, pp. 787—789.

As a result of the supply-demand imbalance, gas companies are not able to supply new residential customers in some cases, and some pipeline companies have had to curtail natural gas deliveries to industrial customers. So far the brunt of the problem has been borne by industrial users.

Growth in Natural Gas Use

The demand for natural gas has grown faster than for any other primary fuel. During the period from the end of World War II until 1970, natural gas use grew at an average annual rate of 6.5%. The growth rate for all other forms of energy used during the same period averaged 3.5% annually. During this period natural gas production increased from 3.3 to 22.0 trillion cubic feet per year (see Figure 6—3). There are some very good reasons for this growth rate, including the following:

1.   Natural gas is the cleanest burning of all primary fuels. No significant sulfur oxide or particulate air pollution results from its combustion.
2.   Natural gas is the most economical fossil fuel. This is due in part to the efficient pipelines used to move it from the production site to the point of consumption. During its rapid

growth period the price of gas was approximately one third
that of crude oil in terms of heating value.
3. Natural gas is the most convenient fuel for domestic use.
Customers do not have to provide fuel storage facilities, and
no specialized burning equipment is required.
4. Natural gas provides valuable raw materials for the petrochemical industry.

Many economists feel that the first two of these reasons are primarily
responsible for the sustained increase in gas consumption.

The nonpolluting nature of natural gas is acknowledged by
industry and government agencies at all levels. Natural gas is practically
free of sulfur and particulate matter, two of the nation's major air
pollutants. The use of natural gas in metropolitan areas can have a
profound impact on air quality. However, according to an Environmental Protection Agency estimate, natural gas consumption would
have to increase by 15% in order to clean up the air of seven major
areas; New York City would have to use one half the total increase.

The following information illustrates the influence natural gas use
has on air pollution levels. During 1970 the coal- and oil-fired electric-
generating plants in one major city emitted 133 thousand tons of sulfur
dioxide while producing 18.5 trillion Btu of useful energy. During the
same time period, a major gas company in the same area emitted
44 tons of sulfur dioxide — mainly the result of added odorants — while
producing 184 trillion Btu of saleable energy. In view of the desirable
characteristics of natural gas, it seems paradoxical that the second major
reason for its high demand is its low price. The cost of home heating
by natural gas in the United States is, on the average, about 30% below
the cost of heating with fuel oil and about 55% below that of electric
heat. Furthermore, natural gas costs most industrial plants and electric
power-generating facilities less than low-sulfur fuel oil or low-sulfur
coal.

Federal Power Commission Price
Controls

A 1954 Supreme Court ruling placed natural gas produced for
sale in interstate commerce under the regulation of the Natural Gas
Act. Since that action was taken, the sale and pricing of natural gas
destined for interstate markets have been subject to control by the
Federal Power Commission (FPC). This agency has allowed some price
increases to be imposed, but it is now generally recognized that natural
gas prices have been maintained at an artificially low level. The average
wellhead price of natural gas, as compiled by the U.S. Bureau of Mines,
from 1956 to 1973, is given in Table 6—8. On the basis of constant

Table 6–8.   NATURAL GAS PRICE TREND

	Cost per Thousand Cubic Feet			Cost per Thousand Cubic Feet	
	Current ¢	Constant ¢*		Current ¢	Constant ¢
1956	10.8	13.3	**1966**	15.7	16.2
1957	11.3	13.4	**1967**	16.0	16.0
1958	11.9	13.8	**1968**	16.4	15.7
1959	12.9	14.8	**1969**	16.7	15.2
1960	14.0	15.8	**1970**	17.1	14.7
1961	15.1	16.9	**1971**	18.2	15.0
1962	15.5	17.1	**1972**	18.6	–
1963	15.8	17.2	**1973**	21.3	–
1964	15.4	16.6			
1965	15.6	16.5			

*1967 is used as the base year.
Data selected from "Trends in Oil and Gas Exploration," Part 2, U.S. Senate Subcommittee Hearings, Aug. 1972, p. 986.

dollars, which accounts for buying power losses to inflation, the average price of natural gas has decreased since 1963.

The results of regulating natural gas prices at such low levels are quite significant and include the following:

1. The exploration and development required to provide new gas supplies have declined. These activities have become economically less attractive because of regulated low selling prices and inflation. The decreasing exploration activity is illustrated in Figure 6–9, and the unfavorable economics are shown by Figure 6–10. The data of Figure 6–10 reflect the fact that the selling price of natural gas has failed to keep pace with increasing costs of well development. The increasing well costs reflect the necessity of drilling to greater depths, drilling in costly locations such as offshore areas, and paying higher prices for labor and materials.

2. The demand for natural gas has been overly stimulated by the artificially low selling prices. This has caused reserves to be depleted at an accelerated rate. Industries, which might otherwise use coal or oil, have switched to natural gas because of the attractive economics as well as the associated advantages related to air pollution problems. This trend has diminished the natural gas supplies available to small residential and commercial consumers.

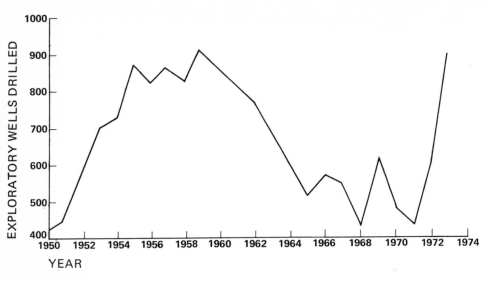

Figure 6–9.   Trend in Natural Gas Exploratory Activity in the Conterminous United States

From "Relationship of Energy and Fuel Shortages to the Nation's Internal Development," U.S. House of Representatives Subcommittee Hearings, Aug. 1972, p. 375. Data for 1972 and 1973 are from *Oil and Gas Journal,* April 22, 1974, p. 152.

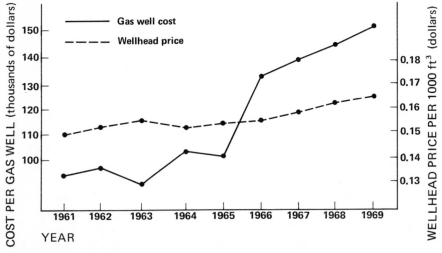

Figure 6–10.   Comparison of Well Cost and Natural Gas Selling Price

From "Relationship of Energy and Fuel Shortages to the Nation's Internal Development," U.S. House of Representatives Subcommittee Hearings, Aug. 1972, p. 379.

3.   Most newly discovered and newly produced natural gas is consumed within the producing state. This practice allows the gas to be sold at higher prices because it does not come under FPC regulations (only the price of gas shipped from one state to another is

regulated). The price of gas produced and sold within a state may reach a free-market level, which is about twice the controlled level. The tendency to selling gas within the producing state creates obvious problems for potential consumers in nonproducing states.

The federal government can deal with the problems and consequences of price regulations in at least three ways. More regulations could be imposed in the form of distribution and production curtailments, rationing, or taxation. Regulations could be diminished and the government could manipulate prices or subsidies as needed. Regulations could be lifted completely and the price of natural gas would seek competitive free-market levels.

Regulation without price changes could be done by reserving natural gas for use by small consumers. These consumers are those who cannot use oil or coal effectively because of cost or the inefficiency involved in the installation and use of millions of small burners and pollution control devices. Since the price would still be low, this approach does not solve the problem of long-term shortages. Remember, this problem results, at least partially, from increasing exploration and development costs coupled with decreasing profits — the costs of finding new gas deposits are too high in relation to the profits obtained.

Drastic price changes have been advocated by some who would double or triple wellhead gas prices. The intention would be to provide incentives for more production and make it uneconomical for large industries and electrical utilities to use gas. This approach involves several risks. Supplies would increase for the small consumer but so would the cost. In addition, greater use of oil by industries would aggravate oil shortages, and increased coal use would be accompanied by air pollution problems.

The third approach, deregulation, would certainly result in higher prices. Market forces (supply and demand) would operate, and the price of gas would arrive at a level competitive with other fuels. It is difficult to predict the extent of the price increase, but it would tend to moderate as industries stopped using gas and switched to other energy sources.

President Nixon, in an April 1973 energy message, recommended the third alternative. This would require the amending of the Natural Gas Act to allow prices paid for gas by interstate consumers to seek levels determined by market forces. The recommended deregulation would be administered in phases to prevent extreme price increases. It would be applied to gas from new wells, gas newly dedicated to interstate markets, and gas supplied despite expiration of contracts. The increased prices of new unregulated gas would be averaged with prices for regulated gas being delivered under existing contracts. The result would be a much more moderate and drawn-out price increase. Such action should stimulate the exploration for new (unregulated) gas supplies.

It should be noted that decreases in exploratory activity are not entirely the results of economic factors. In addition, attractive drilling opportunities are becoming more and more scarce. This latter problem is partially related to federal leasing policies which can prevent access to some promising exploration areas. These problems are similar to those discussed for petroleum in Chapter 5.

Pipeline Imports of Natural Gas

Domestic natural gas production is supplemented by small imported amounts of gas which are piped into the United States from Canada and Mexico. During 1973, these imports amounted to about 1.0 trillion cubic feet. These amounts represented about 4% of the 1973 United States gas consumption. Table 6—9 contains the history of these imports for the 1955—73 period. The amount of gas imported from

Table 6—9.   SUMMARY OF U.S. NATURAL GAS PIPELINE IMPORTS, 1955—73
(billions of cubic feet)

Year	Imports from Canada	Imports from Mexico	Total Imports	Contribution of Total to Domestic Consumption (%)
1955	10.9	0.01	10.9	0.1
1956	10.6	0.01	10.6	0.1
1957	21.1	17.0	38.1	0.3
1958	88.2	46.2	134.4	1.2
1959	81.9	50.9	132.8	1.1
1960	109.9	47.0	156.8	1.3
1961	168.8	51.8	220.6	1.7
1962	342.8	51.1	393.8	2.8
1963	358.0	49.8	407.7	2.8
1964	392.2	52.6	444.9	2.9
1965	404.7	52.0	456.7	2.8
1966	431.9	48.6	480.5	2.8
1967	513.3	51.0	564.3	3.1
1968	604.5	47.4	651.9	3.3
1969	680.1	46.8	727.0	3.5
1970	778.7	41.3	820.0	3.7
1971	910.9	20.7	931.6	4.1
1972	993	—	—	—
1973	1022	—	—	—

Data calculated from "Natural Gas — Salient Statistics," p. 229.2001, Jan. 1974 in *Chemical Economics Handbook,* Stanford Research Institute and "Relationship of Energy and Fuel Shortages to the Nation's Internal Development," U.S. House of Representatives Subcommittee Hearings, Aug. 1972, p. 61.

Canada is seen to have increased steadily — doubling in the 1967—73 period — while the amount received from Mexico has remained essentially constant at about 50 billion cubic feet. The contribution of these imports to the total U.S. consumption has increased only slightly on a percentage basis because the increasing imports have merely kept pace with increasing consumption.

The prospects for increased imports from Mexico are slight. A small proven gas supply and a Mexican policy of energy self-sufficiency indicate that most potential new gas will probably not be available for export. Significant imports from Mexico could end in 1982, the expiration date of current contracts.

Future import increases from Canada may be limited by the actions of the Canadian National Energy Board (NEB). In November of 1971 the NEB turned down three applications for licenses which would have allowed nearly 2.7 trillion cubic feet of gas to be exported to the United States. The applications were rejected on the premise that no surplus gas remained after allowances had been made for reasonably foreseeable Canadian requirements.

Since total pipeline gas imports are not projected to increase much beyond the current 4% of domestic consumption, problems related to a United States dependence on them appear to be small. Most problems which arise from import dependence, as detailed for petroleum in Chapter 5, are not applicable to pipeline imported gas.

Methods for Increasing Natural Gas Supplies

The remainder of this chapter is devoted to a dicussion of three topics that represent possible ways to increase natural gas supplies in the future. These ways are the importation of liquid natural gas, the production of synthetic natural gas, and the nuclear stimulation of gas reservoirs to increase yield.

Liquid Natural Gas

Natural gas is much less dense than solid or liquid fuels and therefore occupies a comparatively large volume per unit of heating value. Because of this, it cannot be transported economically except in pipelines, and it is difficult to store at the sites where it will be used.

These problems can be overcome by converting the gas into a liquid. This change takes place when the gas is cooled to $-259°F$ $(-161°C)$; the volume of the resulting liquid is 600 times smaller than the gaseous volume. This decreased volume makes it economically feasible to store the fuel and transport it to areas not accessible by pipeline. In recent years the manufacture, transportation, and regasification of liquid natural gas (LNG) have become much more common industrial activities.

Liquid natural gas technology has been used commercially in the

United States for more than 30 years. The technology was applied initially to provide supplemental gas during periods of high demand. The use of natural gas by industry is quite steady year-round, but the demand for gas used as a commercial or residential heating fuel increases during the winter. During most winters the demand for heating gas becomes very large for the 10 or 20 coldest days. In some cases these large demands exceed pipeline capacities. When this occurs, LNG that has been previously formed and stored near the point of use is regasified and used to augment the gas arriving by pipeline. More than 30 plants are presently in operation to provide LNG for such uses. Most LNG used during peak demand periods is produced domestically. Foreign LNG was first utilized during the winter of 1968–69 when the LNG equivalent of about 150 million cubic feet of gas was imported from Algeria by the Boston Gas Company.

In addition to solving some high-demand problems, LNG can be used to provide gas during periods of normal use (base-load use). LNG technology offers a way to supplement domestic natural gas supplies with nonpipeline imports, since it makes gas transportation economically feasible between points where pipelines cannot be constructed — across oceans, for example. Some oil-producing countries, such as Algeria and Libya, do not have much local demand for natural gas that is produced along with oil, and also have no large markets accessible by pipeline. A solution to this problem consists of converting the gas into LNG, shipping it by tanker to a consumer area where it can be regasified and delivered by pipeline to customers.

International trade in LNG used for base-load purposes has been going on for about 10 years. It began with the transporting of LNG from Algeria to Britain and France. The United States is just beginning to get involved in LNG importation for base-load use. In March of 1972, the Federal Power Commission gave approval to a plan submitted by the El Paso Natural Gas Company which calls for the importation from Algeria of the LNG equivalent of one billion cubic feet of natural gas per day (0.36 trillion cubic feet/year). Partial initiation of the project is planned for April of 1976, and full production is expected by June 1977. Other projects have also been approved, and the total projected contribution of such activities to the U.S. natural gas supply is given in Table 6–10.

The operations involved in LNG importing projects are much more complex than those associated with similar petroleum projects. Gas liquefaction facilities must be constructed in the exporting country, regasification and storage facilities must be constructed in the importing country, and heavily insulated cryogenic tankers for transoceanic shipment must be provided. The capital investment required to build such facilities is large, a factor that increases the price of LNG and may limit LNG imports.

The total investment required for the previously mentioned El

Table 6—10.   CONTRIBUTION OF LNG IMPORTS TO TOTAL
U.S. NATURAL GAS SUPPLY (trillions of cubic feet)

	1975	1980	1985
Total Natural Gas Supply	24.2	26.9	29.5
Estimated LNG Imports	0.2	0.9	1.6
LNG as Percent of Total	0.8%	3.3%	5.4%

Data selected from "The President's Energy Message and Senate Bill
1570," U.S. Senate Subcommittee Hearing, May 1, 1973, p. 481.

Paso Natural Gas project is estimated to be about $1.7 billion. The
liquefaction facilities in Algeria will cost $350 million, while each of
the required 9 cryogenic tankers will cost an estimated $80—90 million.
Each tanker will be over 900 feet long and have a capacity of 125,000
cubic meters. One cubic meter of liquefied gas is equivalent to 21,900
cubic feet of vaporized gas, so the capacity of each tanker will be
equivalent to 2.7 billion cubic feet of gas. This capacity allows the
projected amount of imported gas to be carried by a total of 135
tanker voyages per year. Two terminal facilities are planned for the
United States — one at Cove Point, Maryland, and the other at
Savannah, Georgia. These regasification facilities are less complex and
expensive than the liquefaction terminals, but each one still represents
an investment of about $100 million. Figure 6—11 shows a general
diagram for a regasification terminal.

The high costs of LNG importation make it necessary that long-
term contracts be established between importers and foreign suppliers.
This practice has some security implications. The export capabilities of
the source country are limited by the number (small because of costs)
and capacities of production facilities. The small number of facilities
makes it much easier for production to be interrupted than is true for
crude petroleum production and export. Little or no flexibility is
available for responding to interruptions of significant duration. Thus,
LNG imports are considered to be less secure than petroleum imports.

The economics of energy supply and demand require that the
delivered price of LNG must be competitive with other fuels after
appropriate allowance is made for special advantages such as low
pollution abatement costs. The estimated costs of LNG are compared
to the costs of other alternative gaseous fuels in Table 6—11. Synthetic
gas and nuclear stimulation are discussed later in this chapter. It is
apparent from the data that LNG is 2—3 times as expensive as pipeline
gas but it is competitive with most other alternatives to pipeline gas.

Some concern has been expressed about potential dangers
associated with transporting huge amounts of LNG by tanker. The
effects of an accidental spill are not considered especially hazardous to

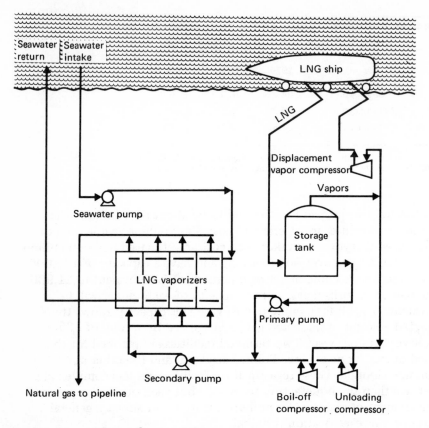

Figure 6–11.   LNG Regasification Facility

Reprinted with permission from The Petroleum Publishing Company from "Design needs for base-load LNG storage, regasification" by Robert N. DiNapoli, *Oil and Gas Journal,* vol. 71, no. 43 (Oct. 22, 1973): Fig. 1, p. 67.

the environment or to people working with the LNG. This is particularly true when LNG is compared to many other acknowledged hazardous products that are transported on a large scale. Recent experimental testing and calculations relative to LNG spills reveal the following:

1. When large amounts of LNG are spilled on water, vaporization takes place rapidly but not instantaneously. The liquefied gas floats on top of the water in a fluid layer that boils and vaporizes at a steady rate. Ice which sometimes forms at the LNG-water interface, reduces the rate of heat transfer to the LNG and slows the evaporation.
2. Sometimes, eruptions occur in LNG that is floating on water. In these eruptions, LNG and water are thrown into the air. The eruptions are believed to be caused by tremendous bubbles which

Table 6—11.   COSTS FOR ALTERNATIVE GAS SOURCES* (dollars per 1000 cubic feet)

	Future price range (1973—80)
Natural gas — associated	0.35—0.95
Natural gas — nonassociated	0.35—0.95
Natural gas — stimulated	0.70—1.50
Synthetic gas — oil or naphtha †	1.00—1.70
Synthetic gas — coal	0.90—1.95
Pipeline gas — from Alaska	1.00—1.35
Pipeline gas — from Canadian Arctic	1.00—1.50
Liquefied natural gas from Alaska**	1.50—2.18
Liquefied natural gas from Canada, Venezuela ††	1.00—2.20
Liquefied natural gas from Eastern Hemisphere	1.00—1.89

* Based on estimates from 7 reporting companies with differing variables.

† Assuming feedstock not exceeding 10 cents per gallon.

** To west coast, United States.

†† Cost varies from Canada and Venezuela, and also depends on point of entry to the United States.

Reprinted from "Fuel and Energy Resources, 1972," Part 2, U.S. House of Representatives Subcommittee Hearings, April 1972, Appendix B, p. 543.

form beneath the surface in superheated LNG; however, they are not explosions.

3. There is no evidence to indicate that LNG ignites spontaneously when spilled onto water or land.

4. Natural gas is less dense than air at the same temperature, and therefore it will not accumulate in low areas. On the contrary, it rises into the atmosphere and is dispersed as soon as its temperature reaches $-160°F$ ($-106.5°C$).

5. The vapors of LNG are not toxic. People can be injured only by breathing vapor in concentrations high enough to cause suffocation from lack of oxygen.

LNG is composed primarily of methane and should not be confused with liquid petroleum gas (LPG) which contains mainly propane and butane. LPG is formed at ordinary temperatures and maintained in liquid form by a moderate increase in pressure. However, natural gas cannot be liquefied by such a procedure. Instead, it is necessary to cool the gas to cryogenic temperatures and maintain these temperatures by insulation as mentioned before.

The increasing availability of LNG has prompted investigations into potential uses of the material as a vehicle fuel. A number of the properties of LNG make such uses attractive. It has an octane rating well above 100 without the addition of antiknock additives, it is clean burning and produces minimal air pollution, its specific energy per pound is 15% greater than gasolines, and it can be used as an engine coolant before it is burned as fuel.

Against these advantages LNG has only a few disadvantages. Its density is about one half that of other common liquid fuels so larger fuel tanks are required, and it must be maintained at low temperatures in heavily insulated fuel tanks. The potential advantages outweigh the disadvantages, as evidenced by the interest expressed in LNG as a potential fuel for rockets, aircraft, locomotives, trucks, and buses.

Synthetic Natural Gas

The technology now exists for converting various petroleum liquids or coal feedstocks into pipeline quality gas which can be used interchangeably with natural gas. The resulting manufactured gas is called *synthetic* (or *substitute*) *natural gas*, SNG. Some processes using certain feedstocks are now commercially feasible, and work is continuing to bring other processes to this point.

The processes for producing SNG are classified according to the physical state of the feedstocks used. Those involving liquids will be our present topic while those in which solid coal is used are dealt with in Chapter 7. More attention is currently focused on commercial coal gasification, since coal is the fossil fuel in greatest domestic supply.

The basic chemistry of gasification, regardless of feedstock, is quite simple. A carbon-containing feedstock is reacted with water at a high temperature to form methane, the principal component of natural gas. However, the overall process requires several steps and is much more complex than our simple chemistry implies.

The liquid feedstock processes that are most highly developed — some are now in commercial use — involve light hydrocarbon feedstocks such as naphtha, natural gas liquids, or liquid petroleum gases. Naphtha is the petroleum fraction with a boiling point between 175° and 240°C. Natural gas liquids and liquid petroleum gases were discussed earlier in this chapter. These liquids are the simplest to work with, since their molecules are less complex and contain fewer carbon atoms than other liquid feedstocks.

The first SNG plants in the United States began production in 1973. In March of that year, a plant located in Harrison, New Jersey, began operations, using naphtha feedstock. This plant, owned by the Public Service Electric and Gas Company, has a capacity of 20 million cubic feet per day. In July 1973 SNG operations began in Marysville, Michigan. It uses natural gas liquids from Canada as feedstock. Other plants have been completed since 1973, and still others are in various stages of construction.

Naphtha has been the most popular feedstock choice for these SNG projects; consequently, the following discussion will emphasize its use. However, the chemistry is similar for processes using natural gas liquids or liquid petroleum gases.

Several different naphtha processes have been developed and commercialized. They are similar in concept but differ in the catalysts used. In a generalized process, vaporized naphtha is superheated under pressure, mixed with hydrogen, and passed over an appropriate desulfurization catalyst. This converts any organic sulfur compounds into H_2S, which is then removed from the system. The process is represented by reactions 6—2 and 6—3.

Hydrodesulfurization

$$R-SH + H_2 \xrightarrow[\text{high T \& P}]{\text{cat.}} RH + H_2S \qquad\qquad 6-2$$

$$R-S-R' + 2H_2 \xrightarrow[\text{high T \& P}]{\text{cat.}} RH + R'H + H_2S \qquad\qquad 6-3$$

The sulfur-free vapor (represented by a tetradecane, a component of naphtha) is then reacted with superheated steam at high temperature ($500°-540°\,C$) and pressure (34 atmospheres) to form synthesis gas. The production of synthesis gas, a mixture of methane, hydrogen, carbon monoxide, and carbon dioxide, is represented by the overall reaction (6—4) which is made up of three steps (reactions 6—5, 6—6, 6—7).

Overall gasification

$$4C_{14}H_{30} + 26H_2O \xrightarrow{\text{high T \& P}} 43CH_4 + 13CO_2 \qquad\qquad 6-4$$

Steam reforming

$$C_{14}H_{30} + 14H_2O \xrightarrow{\text{high T \& P}} 14CO + 29H_2 \qquad\qquad 6-5$$

Hydrogenation

$$C_{14}H_{30} + 13H_2 \xrightarrow{\text{high T \& P}} 14CH_4 \qquad\qquad 6-6$$

Water-gas shift

$$CO + H_2O \xrightarrow{\text{high T \& P}} CO_2 + H_2 \qquad\qquad 6-7$$

More details concerning synthesis gas formation are included in the coal gasification discussion of Chapter 7. The synthesis gas is next subjected to catalytic methanation, a process in which carbon monoxide and dioxide are converted into methane.

Catalytic methanation

$$CO + 3H_2 \xrightarrow{\text{cat.}} CH_4 + H_2O \qquad\qquad 6-8$$

$$CO_2 + 4H_2 \xrightarrow{\text{cat.}} CH_4 + 2H_2O \qquad\qquad 6-9$$

The resulting SNG is purified, dried, and then compressed to pipeline pressure. It contains 95—98% methane and has an energy content of 980—1035 Btu per cubic foot — the same as natural gas.

The most serious problem related to SNG production using light hydrocarbon feedstocks is one of obtaining adequate feedstock supplies. It is unlikely that much surplus domestic naphtha or gas liquids will be available. Therefore, most feedstock must be imported, and the supply becomes subject to the behavior of the producing countries. In addition, the price of naphtha has been escalating rapidly since 1972. This makes the prospects for long-term supply contracts unpromising.

Most naphtha is expected to come from the Caribbean and Europe where there have been some excess supplies. Generally, 9 gallons of naphtha are required to produce 1000 cubic feet of good quality gas (1000 Btu/ft^3). Therefore, about 16.5 million barrels of naphtha per year would be needed as feedstock for a 250-thousand-cubic-feet-per-day SNG plant, assuming the plant operates at 85% of capacity. On the basis of a 10-year supply contract, about 165 million barrels of naphtha would have to be committed to this single large plant. Hope has been expressed that a switch to low-lead or no-lead gasoline production by refineries might increase the supply of naphtha and lower the price.

Heavy liquid hydrocarbons provide an alternative to light feedstocks. Five processes utilizing heavy feedstocks are being investigated, but none are yet ready for commercial use.

The contribution of SNG to future natural gas supplies is estimated in Table 6—12. The figures are arranged to show the fraction of the total SNG production that is expected to come from liquid (petroleum) feedstocks and coal. The liquid feedstock projections assume that the supply problems will be solved.

Table 6—12.   PROJECTED SNG PRODUCTION
(trillions of cubic feet per year)

Year	Feedstock Oil	Coal	Total
1974	0.4	—	0.4
1975	0.7	—	0.7
1980	1.3	0.2	1.5
1985	1.5	0.6	2.1
1990	1.5	2.7	4.2

Data selected from "Trends in Oil and Gas Exploration," Part 2, U.S. Senate Subcommittee Hearings, Aug. 1972, p. 979.

It is expected that the use of liquid feedstocks will be somewhat of a stopgap effort toward solving natural gas shortages. The long-term supplement will likely be SNG produced from coal gasification. Note that in Table 6–12 little growth is projected for liquid feedstock SNG beyond 1980, the date by which it is assumed coal gasification technology will have been perfected.

According to Table 6–11, SNG will cost 2–3 times as much as traditionally obtained natural gas. It is estimated that 70–80% of this price is the result of feedstock costs. Thus, future SNG prices will be very dependent on feedstock prices. In addition, SNG feedstocks create some national security issues since most will have to be imported. Remember, some problems of this type were also found in the case of LNG. A comparison of the different security problems for the two, LNG and SNG, is given below. We see from this comparison that SNG appears to offer considerably more flexibility from a security point of view.

LNG	*SNG*
1. There are few, if any, alternative sources of supply of LNG in the event of interruption of expected supplies.	1. SNG plants may use diverse sources of feedstocks.
2. LNG terminal and regasification facilities are restricted to coastal areas.	2. SNG plants have no geographical restrictions, although dockage facilities are needed to handle feedstocks.
3. LNG imports impose a greater drain on balance of payments, especially in the near term, because capital investments for overseas facilities are large.	3. Imported crude oil feedstocks for SNG avoid foreign processing, thus minimizing balance of payment outflows.
4. Large base-load LNG projects require long lead times for implementation — from 5 to 7 years.	4. SNG plants can be constructed and operated in a relatively short time — from 1.5 to 2 years.
5. LNG does not compete with the petrochemical industry for feedstock supplies.	5. Importing naphtha or other product feedstocks for SNG competes with the petrochemical industry.
6. LNG imports require specialized tankers.	6. Conventional tankers can be utilized to transport SNG feedstock.

LNG   *SNG*

7.   Importing naphtha or other
     product feedstocks leads to
     further exporting of refinery
     capacity. Importing crude
     oil would help restore
     domestic refinery capacity
     and also decrease the
     demand for gas substitutes.

Nuclear Stimulation

Numerous basins containing proven reserves of petroleum and
natural gas are located in the Rocky Mountain belt that passes
through Wyoming, Utah, Colorado, and New Mexico. Some of the
reservoirs within these basins contain substantial quantities of natural
gas which cannot be extracted by conventional technology. The
permeability of the rock making up these so-called tight reservoirs is
very low, and gas contained in the pores of the rock cannot move
easily. The flow of gas to wells drilled into such reservoirs is not fast
enough to allow for economical commercial development.

The permeability and thus the productivity of such reservoirs can
be increased substantially by the detonation of nuclear explosives deep
inside the rock formations. Nuclear explosives are used because they
are small enough to be lowered into a conventional drill hole yet release
huge amounts of energy upon detonation (see Figure 8—5). The
explosion produces a region of broken rock called a chimney and a
fracture system that extends from the chimney into the surrounding
rock. As a result, a stimulated well produces natural gas from an area
with an effective diameter of several hundred feet as opposed to the
6-inch diameter drill hole of a conventional well (see Figure 6—12).

United States Bureau of Mines estimates indicate that 317 trillion
cubic feet of natural gas is potentially available from these Rocky
Mountain deposits if nuclear stimulation techniques are used. The
basins most suitable for such techniques are shown in Figure 6—13.
The estimated potential of 317 trillion cubic feet of gas is based on
stimulated production from only four of the basins that are presently
considered to be most suitable: Uinta, Piceance, Green River, and San
Juan. The significance of the estimated potential (317 trillion cubic feet)
becomes apparent when it is compared to the current proven natural
gas reserves of 250 trillion cubic feet which are now being depleted at
the rate of more than 22 trillion cubic feet annually.

The technique of nuclear stimulation has not been proven
completely feasible and is still in the research and development stage.
Efforts to develop the technology are being conducted, with industrial

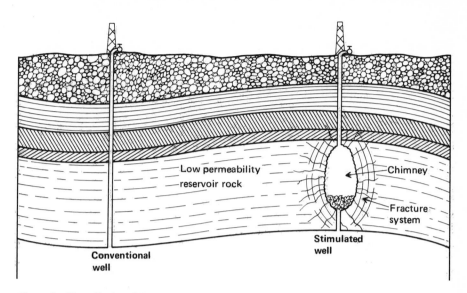

Figure 6—12.   Nuclear Stimulation of Low-Permeability Gas Reservoirs

cooperation, under the Atomic Energy Commission's Plowshare
Program. Three tests had been conducted in the plowshare program
through 1973. The first test, Project Gasbuggy, was carried out in
December of 1967. A 29-kiloton device (equivalent to 29,000 tons of
TNT) was detonated deep below the surface of a site located in
northwestern New Mexico. Project Rulison was the second test and
involved the underground detonation of a 4.3-kiloton device in western
Colorado. The Project Rulison device was detonated in September of
1969. Project Rio Blanco, carried out in May of 1973, was a multiple
explosion type. Phase one consisted of the simultaneous detonation of
three 30-kiloton devices located in a single well at various depths
between 5700 and 6600 feet. A second-stage Rio Blanco experiment,
planned for late 1974 or early 1975, involves the detonation of devices
in 4 to 6 wells on the same day. The third stage of the Rio Blanco
Project will follow the second stage by about 2 years and will consist of
detonations in more than 20 wells on the same day. Project Wagon
Wheel, a new project planned for fiscal year 1974, calls for the sequential
detonation of five 100-kiloton devices buried 11,000 feet under the
surface of the Green River basin of Wyoming. The locations of all these
projects, both completed and planned, are indicated in Figure 6—13.

Results of the completed projects show that nuclear stimulation
can significantly increase gas production. In Figure 6—14, the production
data for the first 17 months of Gasbuggy and the first 9 months of
Rulison are compared to the production of the conventional well
located nearest each test. In 17 months Gasbuggy produced 280 million

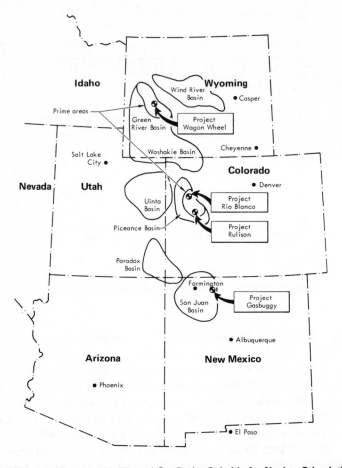

Figure 6–13.   Low-Permeability Natural Gas Basins Suitable for Nuclear Stimulation

Redrawn from "Trends in Oil and Gas Exploration," Part 2, U.S. Senate Subcommittee Hearings, Aug. 1972, p. 572.

cubic feet of gas compared to 70 million cubic feet in 8 years from the nearest conventional well. The results with Rulison are even more dramatic.

If the research and development activities continue to be successful and pilot plant testing reveals no significant problems, commercial development of stimulated natural gas wells could begin by the late 1970s. A proposed schedule of such commercial gas production is given in Table 6–13. This schedule projects an annual production of 1.5 trillion cubic feet by 1985, an amount equal to about 6–7% of current annual domestic production.

One major factor concerning the development and use of nuclear

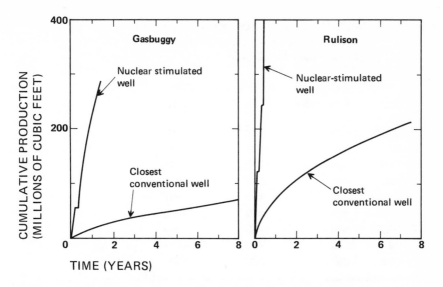

TIME (YEARS)

Figure 6–14.   Gas Production from the Gassbuggy and Rulison Experiments

Redrawn from "Trends in Oil and Gas Exploration," Part 2, U.S. Senate Subcommittee Hearings, Aug. 1972, Fig. 8, p. 575.

stimulation technology is public acceptance. The Gasbuggy experiment generated no formal opposition. However, several injunctions were sought to stop the detonation of the Rulison Project. The courts ruled that the experiment could proceed. Rio Blanco generated even more opposition, but the courts again allowed the experiments to proceed.

The geographical areas suitable for nuclear stimulation development are now or have been used for exploratory and developmental work related to petroleum and natural gas production. Nuclear stimulation of such areas involves only two effects that are not common to conventional petroleum or gas field development. These effects are residual radioactivity and seismic waves resulting from the nuclear explosion.

Data collected from the Gasbuggy and Rulison experiments showed that 3 months after detonation the only radioactive isotopes of significance remaining in the resulting gas were tritium (hydrogen-3), krypton-85, and carbon-14. The amount of carbon-14 produced is so slight that its potential contribution to a radiation hazard is small compared to that of the tritium. All of the krypton remained in an uncombined state while most tritium was incorporated into hydrogen-containing compounds — primarily water and methane. High early-production rates cause most of the radioactive contaminants to be removed during the first year of production. The second year, average radioisotope concentration is about 1000 times lower than that of the first year. Table 6–14 shows how the radiation dose resulting from

Table 6—13.   PROPOSED SCHEDULE OF COMMERCIAL GAS
PRODUCTION BY NUCLEAR STIMULATION
(billions of cubic feet per year)

Year	Wells in Production	First-year Gas	Total Gas
1977	20	85	85
1978	50	133	178
1979	110	272	381
1980	190	369	607
1981	290	465	867
1982	390	465	1076
1983	490	465	1249
1984	590	465	1424
1985	690	465	1590

Data selected from "Trends in Oil and Gas Exploration," Part 2,
U.S. Senate Subcommittee Hearings, Aug. 1972, p. 582.

stimulated gas use compares to the dose from other selected sources.
California is assumed to be the market area. Note the different
exposures for stimulated gas use which depend on the uses made of
first-year gas. When it is used only in power generating plants, the
maximum exposure would result in locations downwind from the

Table 6—14.   RADIATION EXPOSURE TO THE POPULATION OF
CALIFORNIA FROM VARIOUS SOURCES (millirems per year)

Source	Estimated Individual Exposure — 1970	Estimated Individual Exposure — 2000 Average	Maximum
Natural background radiation	114	114	—
All medical sources	98	160	—
Nuclear atmospheric tests	5	5	—
Nuclear power reactors	0.002	~0.2	—
Power-reactor fuel reprocessing	0.0008	0.2	—
Gas stimulation general use of all gas	—	<0.45	<0.7
power plant use of first-year gas	—	<0.11	<2.1

Data selected from "Trends in Oil and Gas Exploration," Part 2, U.S. Senate Subcommittee
Hearings, Aug. 1972, p. 596.

plants. However, even this dose is much lower than that from normal background radiation.

Seismic shock waves result when large underground nuclear (or conventional) explosions take place. These shock waves can cause damage to structures located in the vicinity of the explosion. The number of structures exposed to such hazards is low in the basins under consideration for nuclear stimulation. Also, it is anticipated that well sites will be located such that any damage which does occur will be slight and repair costs will not be excessive. In order to minimize the number of days to which local populations would be subjected to this seismic nuisance, a number of wells could be prepared ahead and detonated on the same day. Four detonation days should suffice for a field operation involving 30 to 40 wells per year.

The problems of radioactive gas and seismic shock waves could be eliminated by the development of other stimulation techniques. In 1974 the Atomic Energy Commission and the Department of the Interior contracted with industrial interests to partially finance a project to test the feasibility of using hydraulic pressure fracturing for gas well stimulation. Hydraulic pressure fracturing involves injection of a fluid under high pressure into a well to create cracks in underlying rock.

Suggested Readings

Crouch, W. W., and J. C. Hillyer, "What Happens When LNG Spills?" *Chemical Technology*, April 1972, pp. 210–215.

De Nevers, N., "Liquid Natural Gas," *Scientific American*, Oct. 1967, pp. 30–37.

DiNapoli, R. N., "Design Needs for Base-Load LNG Storage, Regasification," *Oil and Gas Journal*, Oct. 22, 1973, pp. 67–70.

Hardy, E. F., "The Emergence of U.S. Gas Utilities as a Factor in World Petroleum Economics," *American Gas Association Monthly*, May 1974, pp. 7–11.

Kridner, K., "LNG in the Energy Industry," *Energy Pipelines and Systems*, July 1974, pp. 34–35.

Maugh, T. H., II, "Gasification: A Rediscovered Source of Clean Fuel," *Science*, Oct. 6, 1972, pp. 44–45.

"Nuclear Stimulation of Natural Gas," Hearings before the U.S. Senate Subcommittee on Public Lands, May 1973, Government Printing Office, Washington, D.C.

"Synthetic Natural Gas — A Special Report," *Oil and Gas Journal*, June 25, 1973, pp. 107–122.

Chapter 7
Coal

Coal began to be used as an energy source in the United States in 1830, and by 1850 it satisfied slightly more than 10% of the nation's total energy requirements. This increased to approximately 35% by 1875 and to 70% by 1900. By 1925 coal was unchallenged as the major fuel in the United States and was still the source of approximately 70% of the total energy used. During the next 25 years both oil and natural gas replaced coal in many applications, and by 1950 coal was providing only 37% of the energy used in this country. The decline has continued and coal now satisfies only about 20% of the national energy requirements. Refer back to Figure 1—3 for a graphical representation of the changing role of coal in the United States energy use, and to Figure 3—3 for an estimate of its current contribution.

Coal is not an obsolete fuel today in spite of its shrinking percentage contribution to the total energy used. Energy use today is much greater than it was in the past; much more coal is needed to provide 20% of our present energy than was used in 1900 when coal provided 70%. The use of coal is projected to increase in the future (see Figure 1—4). This projection is based, at least in part, on the fact that coal reserves are known to be much larger than those of any other fossil fuel. However, many problems related to the use of coal remain to be solved. For this reason, a great deal of research is now aimed at developing methods for using these vast reserves in ways that are compatible with environmental considerations. Coal will be an important part of future energy production.

Origin of Coal

Coal is the product of the action of various chemical and physical processes on buried, partially decomposed vegetation. The plant origin of coal is verified by fossil remains and stratification found in coal beds.

Partially decomposed vegetation suitable for coal formation is obtained only under specific conditions. Accumulated plant material is usually in contact with the atmosphere, or it is found in highly aerated water. Under these conditions, decomposition by fungi and bacteria is continuous and complete. It is only under anaerobic (oxygen-free) conditions that partial decomposition takes place. Such conditions often prevail in stagnant swamp water where fallen vegetation is soon covered by other debris. Under these conditions the plant protoplasm, proteins, starches, and, to a lesser degree, cellulose are decomposed by anaerobic bacteria. Other plant components such as lignin (a cellulose glue) and protective waxes are much more resistant to anaerobic bacterial action and remain essentially unchanged.

Heat and pressure, acting for long periods of time, change the chemical composition of the partially decomposed plant material into that characteristic of coal. The heat and pressure result primarily from the steady accumulation of overburden, which is composed of more recent sediments and debris.

Coal is generally classified according to the degree of increased exposure to the metamorphic conditions of heat and pressure. Four broad sequential stages of coal formation are recognized; each stage represents an increase in the carbonization (percent carbon) of the original plant material. The degree of carbonization is known as the *rank* of the coal. The various ranks of coal are given in Table 7—1 along with some characteristics of each rank.

It is generally agreed that the pressures and temperatures resulting from increasing overburden are sufficient to cause carbonization only to the rank of bituminous coal. Further increase in rank to anthracite requires the higher temperature and pressure reached only during geologic, mountain-building processes.

The natural coal-forming process has been partially duplicated in laboratory experiments. In the late 1930s, the German fuel chemist Ernst Berl succeeded in making synthetic coal. He subjected woody vegetable matter to temperatures and pressures similar to those produced by rocky overburden. The result of his experiment was a material quite similar to lignite. Since then, coals with ranks up through bituminous have been made in laboratories. However, the process has no immediate practical significance since more energy is required to produce the coal than can be recovered by burning it.

Chemical Composition and Structure of Coal

The chemical composition of coal is affected by differences in the incorporated plant materials, the degree of carbonization that has occurred, and the impurities absorbed from the surroundings during the formation process. The approximate compositions observed for various ranks of coal are given in Table 7—2. Note that the compositions are given on a dry, mineral-free basis. In addition, the normal moisture content and the heat yielded upon combustion (calorific values) are included. The two major changes in composition that occur

Table 7—1.   STAGES IN COAL FORMATION

	Peat	Partially decayed plant material found accumulated in swamps and bogs. Not a true coal since it has not been subjected to heat and pressure. Little or no carbonization has occurred. Considered to be the first stage of coal formation. Low heat value. Nonetheless dried and used as fuel in some countries.
	Heat and pressure ↓	
	Lignite	First stage in the carbonization of peat. Also known as *brown coal*. Contains quite a bit of woody structure. Colored dark brown to black. Limited use as a fuel because of low heat value.
Increasing carbon content ↓ and rank	Heat and pressure ↓	
	Bituminous coal (soft coal)	Woody structure has disappeared. Used extensively as an industrial fuel and as a source of coke. Burns with a smoky, sooty flame.
	Very high temperature and pressure ↓	
	Anthracite (hard coal)	Final stage of coal formation. Burns with a clean flame. Formerly used as a domestic fuel but now largely replaced by natural gas.

as the carbonization process goes on are readily apparent: the percentage of carbon increases and the percentage of oxygen decreases.

Coal contains many other elements in addition to those given in Table 7—2. Most of these elements are present only in trace amounts. Some were contained in the original plants while others were incorporated during the carbonization process. Those in the latter category came from minerals deposited at the time of peat formation or from secondary minerals that crystallized from water percolating through coal seams.

Table 7–2.   THE CHEMICAL COMPOSITION OF COAL

Type of Coal	Carbon* (%)	Hydrogen* (%)	Oxygen* (%)	Nitrogen* (%)	Moisture as Found (%)	Calorific Value* (Btu/lb)
Peat	45–60	3.5–6.8	20–45	0.75–3.0	70–90	7,500–9,600
Brown coal and Lignites	60–75	4.5–5.5	17–35	0.75–2.1	30–50	12,000–13,000
Bituminous coals	75–92	4.0–5.6	3.0–20	0.75–2.0	1.0–20	12,600–16,000
Anthracites	92–95	2.9–4.0	2.0–3.0	0.50–2.0	1.5–3.5	15,400–16,000

*Dry, mineral-free basis.

Modified with permission from *Encyclopedia of Chemical Technology*, 2nd ed., by R. E. Kirk and D. F. Othmer (New York: John Wiley and Sons, Inc., 1964), vol. 5, p. 627.

The elemental composition of coal, particularly the amount of sulfur and ash, is important when the environmental impact of coal burning is considered. The problem of sulfur in coal is dealt with in detail later in this chapter. Coal ash is the solid residue remaining after the coal is burned. Ash should not be confused with mineral matter. Although the two are related, the amount of ash can be more than, equal to, or less than the amount of mineral matter in the coal. The amount of ash depends upon the minerals present and the conditions under which combustion takes place. Most coal contains between 5% and 10% ash. The major constituents of ash are given in Table 7–3, but it also contains many minor components not included in the table. Ash represents an impurity which increases coal transportation

Table 7–3.   THE CHEMICAL COMPOSITION OF
INORGANIC ASH

Chemical Substance	Range of Weight Percent
Silica (SiO_2)	34–38
Alumina (Al_2O_3)	17–31
Iron oxide (Fe_2O_3)	6–26
Calcium oxide (CaO)	1–10
Magnesium oxide (MgO)	0.5–2
Sulfur trioxide (SO_3)	0.2–4
Unburned carbon (C)	1.5–20

Reprinted with permission from L. J. Minick, *American Society for Testing and Materials Proceedings*, vol. 54, no. 1129 (Philadelphia: American Society for Testing and Materials, 1954): 1129.

costs and creates a disposal problem after coal is used. Under some conditions ash also contributes to air pollution and the corrosion of burning equipment.

Coal is chemically a complex mixture of organic compounds which behaves in a manner unlike that of any other class of chemical materials. In spite of many investigations, the molecular structure is not fully understood. However, it is generally agreed that the basic molecular skeleton of coal is made up of various arrangements of the six-carbon ring of aromatic hydrocarbons (see Figure 7—1A). The aromatic rings may be fused with other rings to form discrete clusters (see Figure 7—1B). It is considered likely that most of the carbon in coal is in the form of condensed-ring clusters with the clusters arranged in a fairly ordered pattern. The outer rings of the clusters are bonded to hydrogen atoms or other chemical groups that readily replace hydrogen atoms. In addition, some atoms of nitrogen, sulfur, and various metals are found in some of the rings (in place of carbon atoms) or between clusters.

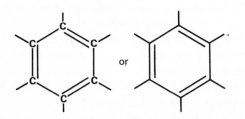

A. FUNDAMENTAL AROMATIC UNIT OF COAL

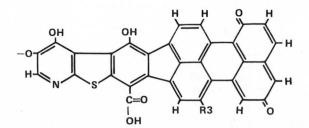

B. CLUSTER OR FUSED RING UNIT REPRESENTATIVE
OF THOSE THOUGHT TO BE FOUND IN COAL

Figure 7—1.   The Molecular Structure of Coal

Figure B modified with permission from "Some Aspects of Coal Research" by George R. Hill, *Chemical Technology* (May 1972): part of Fig. 3, p. 294. Copyright by the American Chemical Society.

Coal Resources and Reserves

Coal is the most abundant fuel found in the United States. The increasing shortage of energy sources makes our domestic coal reserves an asset of significant potential value. The practical use of these reserves is related to the following questions:

1. What is the extent of U.S. coal reserves as determined by exploration and mapping?
2. What percentage of the coal in these reserves can be removed with available mining technology?
3. Where are the reserves located geographically?

The total remaining coal in reserve in the U.S. as of December 31, 1970, was estimated to be 1576 billion tons. These reserves consist of 1113 billion tons of bituminous coal, 447 billion tons of lignite, and 16 billion tons of anthracite. Approximately 50% of this total, or about 788 billion tons, is considered to be unmineable because the coal seams are too thin, are buried too deep, or underlie surface features that preclude mining. Only about 532 billion tons of the remaining 50% will actually be recovered. This is based on an assumed recovery of 57% by underground mining and 80% by surface or strip mining. Therefore, only about 34% of the estimated total U.S. reserve can be considered recoverable with available mining technology. The recoverable amount of each rank of coal is given in Table 7—4. (Although 1970 data are used, they are still good estimates of U.S. coal reserves, since total production has been only about 0.6 billion tons per year since 1971.)

Figure 7—2 shows the geographical distribution of coal fields in the United States. The coal deposits are obviously very widely distributed across the country. However, the rank of the coal is not evenly distributed among all the deposits. This is indicated in Figure 7—3 which shows the states that contain the largest reserves of each of the basic coal ranks. Note that a single state contains a dominant part of the reserves for two of the three ranks. Pennsylvania contains essentially

Table 7—4.   RECOVERABLE COAL RESERVES IN THE UNITED STATES, Dec. 31, 1970.

Coal Rank	Estimated Reserves (billions of tons)	Estimated Recoverable Reserves (billions of tons)	Estimated % Recoverable
Lignite	447	179	40.0
Bituminous	1113	360	32.3
Anthracite	16	3	18.8
Total	1576	542	34.4

Data from "United States Coal Resources and Production," U.S. Dept. of the Interior, Bureau of Mines, June 1971, pp. 3, 9, 10.

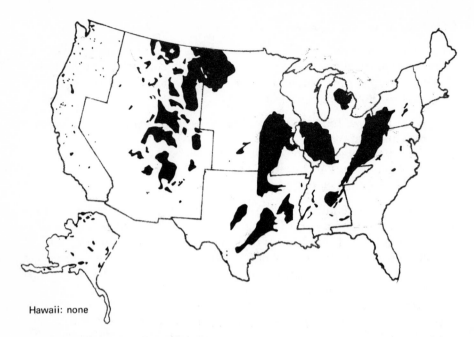

Figure 7—2.   Coal Fields of the United States

From *United States Energy Fact Sheets, 1971,* U.S. Dept. of the Interior, Feb. 1973, p. 129.

all (94.6%) of the anthracite, while North Dakota is the dominant lignite source with Montana a weak second. Bituminous coal reserves are more widely and evenly distributed.

The heat content of bituminous coal varies considerably (see Table 7—2). Coal with heat content at the lower end of the bituminous range is often referred to as subbituminous coal. This subclassification is used in Figure 7—3. Slightly more than one third of the total bituminous coal reserves are in the subbituminous category. An analysis of the distribution of bituminous and subbituminous reserves between various states reveals an important fact: most bituminous coal is located east of the Mississippi River while all of the subbituminous is found west of the Mississippi. The importance of this fact will become apparent later in this chapter when some of the problems related to the use of coal are discussed.

Coal Production

The annual production of coal in the United States by rank for the 1930—73 period is given in Figure 7—4. Note that the production is given on a logarithmic scale. We see that the total U.S. production in 1973 amounted to 598.5 million tons. This total was composed of

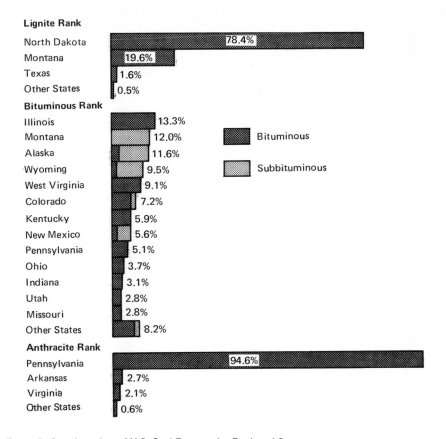

Figure 7—3. Location of U.S. Coal Reserves by Rank and State

Data from "United States Coal Resources and Production," U.S. Dept. of the Interior, Bureau of Mines, June 1971, pp. 4–8.

584.8 million tons of bituminous, 6.8 million tons of anthracite, and 6.9 million tons of lignite. It is apparent that bituminous rank coal makes up most of the total U.S. production — 97.7% of the 1973 total. Note that the bituminous production line and the total line nearly converge in Figure 7—4.

The significance of the amount of coal in the reserves discussed previously does not become clear until it is compared to current production figures. A ratio of recoverable reserves to annual production gives us some idea of the potential lifetime of the reserves. For 1970, with a total production of 606.0 million tons and an estimated recoverable reserve of 532,000 million tons (from Table 7—4), this ratio is 878 to 1. Perhaps a more meaningful comparison is given by the ratio of recoverable bituminous reserves to annual bituminous production, since nearly all of the production in the U.S. involves coal of bituminous rank. This ratio (350,000 to 590.6) is equal to 593 to 1. The present rate of

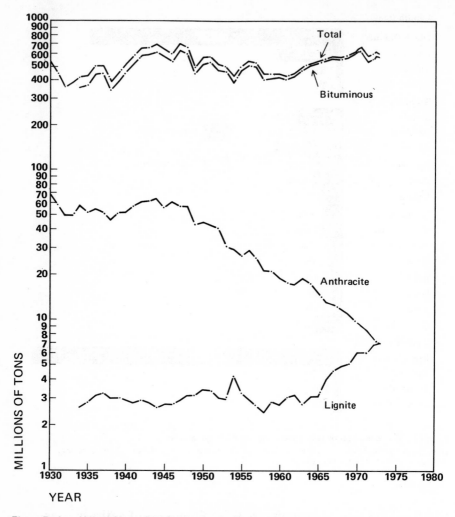

Figure 7—4.   United States Coal Production by Rank

Redrawn from "Coal — Production by Type," *Chemical Economics Handbook*, p. 229.6530, Sept. 1971, Stanford Research Institute.

production will therefore decrease the coal reserves only slightly each year since consumption is small compared to availability. Even if coal use were to double or triple, the reserve would still represent a sizeable energy source for the next century or two.

In Figure 7—5 the 1970 bituminous and lignite coal production is shown for each state. Although there was coal production in 24 states, most of it was concentrated in a small region of the country. Five states accounted for 78% of the total production. They are, in order of decreasing production: West Virginia, Kentucky, Pennsylvania, Illinois,

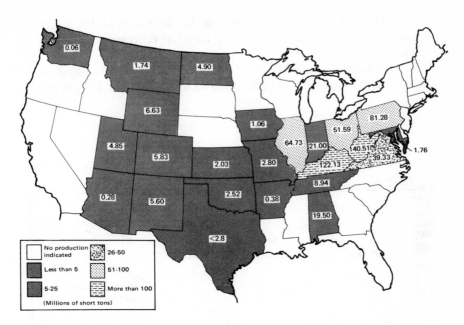

Figure 7—5.   Production of Bituminous Coal and Lignite in the Conterminous United
States, 1970

From "United States Coal Resources and Production," U.S. Dept. of the Interior, Bureau of
Mines, June 1971, p. 19.

and Ohio. It is also worth noting that 93% of the production took place
in states east of the Mississippi River.

Coal Use

The pattern of coal consumption in the United States has under-
gone dramatic changes with time as shown by the data of Figure 7—6.
This is especially true for the 30-year period following the end of World
War II. In 1945 the four major coal consumers were railroads, retail
deliveries (home heating, etc.), coke production, and electric power
utilities. Each one consumed approximately the same amount of coal
annually, and their total consumption amounted to nearly 75% of the
national total. Since that time, the following marked changes have
occurred:

1.   Railroad consumption has decreased to an amount representing
less than 2% of the 1945 level. The diesel engine has replaced the
coal-fired steam engine.
2.   Retail deliveries have decreased to less than one tenth the level of
the postwar high. Fuel oil and natural gas have largely replaced
coal as a home heating fuel.
3.   Electric power utility consumption has increased to over 5 times

the 1945 level. This one area is now the dominant factor in U.S. coal consumption, with an annual use that represents 70% of the national total.

4. The amount of coal consumed in coke production has fluctuated above and below the 1945 level by as much as 20%. Today, how-

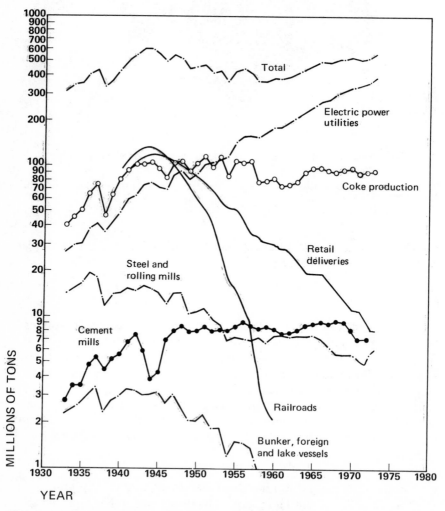

*Data include both bituminous coal and lignite.

Figure 7–6.   Bituminous Coal Consumption in the United States*

Redrawn from "Bituminous Coal — Consumption," *Chemical Economics Handbook,* p. 212.1150, Sept. 1971, Stanford Research Institute.

ever, the level is about what it was in 1945. Coke production is the second largest market for coal at present, accounting for 17% of the 1973 total production.

Coal and Electric Power Production

As mentioned above, the production of electricity represents the primary use of coal in the United States. The essential features of a coal-burning power plant are quite simple. Coal is pulverized and blown into giant furnaces where combustion takes place. The main chemical reaction that occurs is:

$$C + O_2 \longrightarrow CO_2 + heat \qquad\qquad 7\text{--}1$$

The hot exhaust gases are passed over water-containing pipes designed to absorb heat and generate steam which in turn drives turbine electric generators. A schematic diagram of such a power plant is given in Figure 7—7.

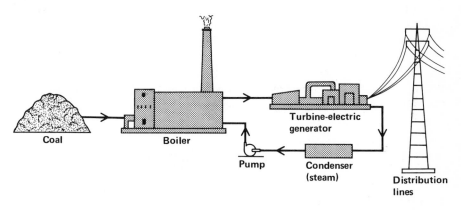

Figure 7—7.   Coal-Burning Power Plant

Modified by permission of the publisher and author from *Introduction to Natural Science, Part 1: The Physical Sciences* by V. L. Parsegian, et al. (New York: Academic Press, 1968).

Coke Production from Coal

The second largest use of coal occurs in the production of coke, used primarily in steel making. Coke is produced when bituminous coal is heated to a temperature of $1000°\text{--}1300°C$ in the absence of air. Volatile materials are driven off and elemental carbon in the form of coke is left as a residue. A part of the volatile material that distills off is condensed into a liquid. The liquid is made up of a water layer and an oily layer. The water layer, called *ammonia liquor*, is used for the subsequent production of ammonium sulfate fertilizer. The oily layer contains coal tar and a crude, light oil. The remainder of the volatile distillate is called *coal gas*. About 70% of the coal is converted to coke in the coking operation; 6—10% becomes coal tar, light oil, and ammonia liquor; and the remainder is coal gas. Table 7—5 contains a list of a number of the products that distill off during the coking operation.

Coke is a very good fuel, but it is too expensive at present for most industrial uses except in blast furnace operation. In a blast furnace coke becomes a chemical raw material as well as a fuel. Carbon monoxide (CO), a gaseous product of the burning coke, reacts with iron oxides in a blast furnace and produces free elemental iron.

Coal tar is a valuable substance because it contains 65—90% aromatic compounds and some heterocyclic compounds. Aromatics are organic compounds that contain a characteristic cyclic carbon structure (see Figure 7—1A). Heterocyclics are cyclic organic compounds in which at least one noncarbon atom is included in the ring structure. Plants and animals produce aromatic and heterocyclic compounds, but only in very small quantities. Coal tar is consequently the richest natural source of these compounds.

Many useful products are manufactured from the aromatic and heterocyclic compounds of coal tar. These products include dyes and coloring agents, plastics, pesticides, explosives, textiles, preservatives, drugs, and other medicines. Therefore, in addition to being an important energy source, coal is valuable to the chemical industry as a source of raw materials. Some people have even suggested that coal is such a valuable source of chemicals, its extensive use as a fuel should be discouraged.

Table 7—5.   DISTILLATION PRODUCTS OF THE COKING PROCESS

Name	Pound(s) per Ton of Coke	Name	Pound(s) per Ton of Coke
Coal tar and light oil		**Coal gas**	
Benzene	11.80	Carbon monoxide	43.2
Toluene	2.72	Hydrogen	30.4
Xylenes	1.33	Methane	132.0
Naphthalene	6.48	Hydrogen sulfide	6.7
Phenanthrene	2.26	Hydrogen cyanide	1.7
Anthracene	0.64	Ethylene	19.6
Pyridine	0.16	Propylene	3.4
Quinoline	0.13	**Ammonia liquor**	
Phenol	0.95	Ammonium sulfate	20.0

Reprinted by permission of the publisher, from Curtis B. Anderson, Peter C. Ford, and John H. Kennedy, *Chemistry: Principles and Applications* (Lexington, Mass.: D.C. Heath and Company, 1973), p. 149.

Coal gas (coke oven gas) is used primarily as a fuel gas in the operation of steel plants with which most coke-producing facilities are associated.

Distribution of Coal Use

When the primary uses of coal are considered, it is not surprising to find an uneven geographical distribution of coal consumption. Most coal is generally used in those parts of the nation with high population densities (and high demand for electricity) or in parts of the nation containing large steelmaking operations, as shown in Figure 7—8.

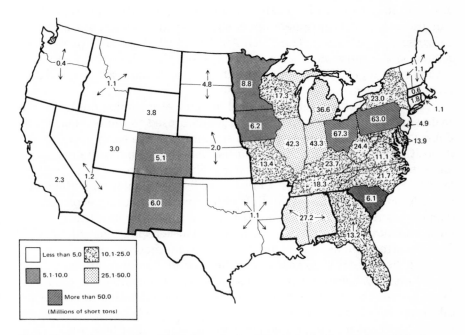

Figure 7—8.   Consumption of Bituminous Coal and Lignite in the Conterminous United States, 1970

From "United States Coal Resources and Production," U.S. Dept. of the Interior, Bureau of Mines, June 1971, p. 26.

Consumption is highest in Ohio and Pennsylvania — each state consumed more than 60 million tons in 1970. Michigan, Indiana, and Illinois each used between 26 and 50 million tons. The less populated western states used very little as did the heavily populated West Coast. The surprisingly low consumption on the West Coast is probably due, at least in part, to a poor coal supply (see Figure 7—2).

Problems Related to Coal Use

Domestic reserves of coal are large in both an absolute sense and when compared to current annual consumption figures. Despite this, a genuine concern has developed about using all recoverable coal reserves to help alleviate energy shortages. This concern is caused primarily by problems related to the mining and use of coal as a fuel. The problems are mainly in three areas: (1) air pollution, (2) environmental impact of mining methods, and (3) the solid form of coal at normal temperatures.

The chief objections to the use of coal from an air pollution standpoint are its sulfur and ash content. All coals contain some sulfur as a normal impurity. Upon combustion, this sulfur is converted into gaseous sulfur oxides which can be emitted into the atmosphere. Ash can also enter the atmosphere under some conditions and contribute to particulate pollution. The Clean Air Act of 1970 contains provisions related to the emission of such pollutants, but there is concern about whether these provisions can be met if coal use is increased to alleviate energy shortages.

Another problem related to coal use concerns mining methods. The main concern involves strip mining, a method used to remove coal from deposits located near the surface. In strip mining, the earth covering a deposit is removed by machines and the underlying coal is then easily and safely obtained. In the past, water pollution, decreased soil fertility, landslides, and landscape destruction have been the environmental and social costs of using the method. A desire to prevent future ecological damage has resulted in numerous regulations aimed at preventing environmental disruption by strip mining. These regulations will profoundly affect the use and economics of this method of coal mining.

At the present time coal cannot be used in a number of major energy markets because it is not in a physical form compatible with existing equipment — automobile or aircraft engines and home gas or oil furnaces, for example. If coal is to be used in such markets, methods for converting it into the liquid and gaseous states must become economically feasible. A large amount of research is now devoted to these tasks.

Each of these three problem areas briefly mentioned here will be discussed in detail in the remaining sections of this chapter.

Sulfur in Coal

The presence of sulfur in coal is not surprising when the origin of coal is remembered. Plant life and natural waters are both necessary for coal formation and both contain sulfur as a normal constituent. Some of this sulfur survives the carbonization process and remains in the resulting coal. The actual amount present depends upon the type of plants originally involved and the conditions under which carbonization proceeded. The sulfur content of coal ranges from less than 1% to about 7% by weight. Figure 7—9 shows the average sulfur content of coal

Figure 7—9.   Average Sulfur Content of Coal Deposits in the Conterminous United States (Percent)

From "Average Sulfur Content of U.S. Coal Deposits," J. Calvin Giddings, *Chemistry, Man and Environmental Change* (New York: Harper & Row, 1973), Fig. 7–12, p. 262.

reserves in states having substantial deposits. Since coal is not uniform in composition, some in each state can be found with a sulfur content less than the average shown.

Sulfur in coal becomes a problem when the coal is burned. During the combustion process any sulfur in the coal is converted to gaseous sulfur oxides — mainly sulfur dioxide (SO_2) and a small amount of sulfur trioxide (SO_3). The reactions are:

$$S + O_2 \longrightarrow SO_2 \qquad \qquad 7–2$$

$$2SO_2 + O_2 \longrightarrow 2SO_3 \qquad \qquad 7–3$$

Both of these gases are air pollutants and excessive amounts have been shown to be hazardous to both human health and plant life. Consequently, emission standards have been established and must be met if we continue using coal as a fuel.

Three approaches to the problem of decreasing sulfur oxide emissions from coal-burning processes are:

1. Use only coal with an original sulfur content low enough to meet the emission standards.
2. Prior to combustion, remove enough sulfur from high-sulfur coal to meet the emission standards.
3. Burn high-sulfur coal and remove the sulfur oxides from the exhaust gases before releasing them into the atmosphere.

Current investigations reveal that each approach has advantages and disadvantages.

If no coal pretreatment or exhaust gas cleanup is used, then coal burned in new power plants will have to contain no more than 0.7% sulfur and have a high heat content in order to meet emission standards. The standards to be met are based on sulfur emissions per unit of heat obtained from the coal. For new large power plants the standard is 0.6 pounds of sulfur for each million Btu of heat obtained. When low-sulfur coal is available, its use offers a simple solution to meeting the emission standards. The result of an analysis of the sulfur content of known domestic coal reserves is given in Table 7—6.

At first glance the numbers of Table 7—6 make the use of low-sulfur coal appear promising. It must be remembered, however, that these data include all ranks of coal, and the geographical location of the reserves is not specified. Both of these factors are against the use of low-sulfur coal in large amounts. Figure 7—10 takes into account the location of U.S. coal reserves. We see from these data that the majority of low-sulfur coal reserves are located in the western United States and are therefore far removed from the major eastern consumers (see Figure 7—8). Coal production in the western states is still quite limited and amounts to only about 5% of the national total. Even if production were higher, transportation costs to the east would be prohibitively high.

Table 7—6.   SULFUR CONTENT OF TOTAL REMAINING
U.S. COAL RESERVES

Sulfur Content (%)	Percent of Total Recoverable Reserves
0.7 or less	46
0.7—1.0	19
1.0—3.0	15
Greater than 3.0	20

Reprinted with permission from *U.S. Energy Outlook* (Washington, D.C.: National Petroleum Council, 1972), p. 160.

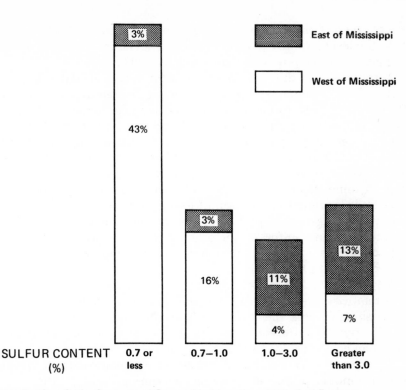

Figure 7—10.   Sulfur Content and Geographical Location of U.S. Coal Reserves

Redrawn with permission from *U.S. Energy Outlook* (Washington, D.C.: National Petroleum Council, 1972), p. 160.

The rank of the low-sulfur coal also presents a problem. The bulk of the western reserves are of the lignite rank and have characteristically low heat content. As the heat content of coal decreases, the sulfur content must also decrease if emission standards are to be met. Eastern coal containing 0.72% sulfur and yielding 24 million Btu per ton is equivalent to western coal which yields 16 million Btu per ton and has a sulfur content of only 0.48%. Thus, much of the low-sulfur western lignite still exceeds performance standards in terms of sulfur emissions per unit of heat produced.

Because of these problems, the use of low-sulfur coal does not represent a universal solution to the problem of sulfur oxide emissions. At best it provides a solution only in limited areas of the country that have deposits of suitable coal.

The cleanup (desulfurization) of high-sulfur coal prior to its use has been studied by both governmental and private organizations for many years. Two basic approaches to the problem have been followed, physical cleaning and chemical cleaning. Limited technology has been developed and is now used to physically remove sulfur from coal.

However, the technique is not effective in reducing the sulfur in most coals to the level required to meet emission standards. Attempts are still being made to develop new and more effective processes.

A knowledge of the form in which sulfur occurs in coal is an important part of understanding the various sulfur removal techniques. Sulfur occurs in three forms: inorganic pyritic sulfur, organic sulfur, and inorganic sulfates. The sulfates are usually present only in small amounts and are not generally regarded as a problem. On the average, the sulfur content of coal is about equally distributed between pyritic and organic sulfur. However, some exceptions to this rule do occur, and the sulfur content of some major eastern U.S. coal reserves is mainly pyritic. Pyritic sulfur is present in the form of iron pyrite, FeS_2, and is generally found as discrete pyrite particles rather than as a part of the coal matrix. Organic sulfur, on the other hand, occurs as part of the organic molecules making up the coal matrix and is therefore chemically bonded to atoms of the matrix.

A type of physical coal cleaning called *coal washing* is now widely used to remove some pyritic sulfur as well as ash and shale from coal. In the washing process, the coal is finely ground and then washed with water. The pyritic sulfur with a specific gravity of about 5.0 settles more rapidly than coal with its specific gravity of 1.2—1.7. In this way some small amounts of pyritic sulfur are separated and removed from the coal. The actual amount of iron sulfides removed during washing varies greatly between different coals. Chemically bound organic sulfur is affected very little by physical cleaning processes and in general cannot be removed in this way. Physical cleaning lowers the sulfur content of coal but usually only by a small amount. For this reason it is only useful for cleaning those coals with sulfur contents just above the allowable limit. In many cases the cost is too high to justify the slight decrease in sulfur level that is realized.

A great deal of research is under way with the goal of developing practical chemical processes for cleaning coal. We will discuss three processes that are still experimental in order to illustrate the type of work going on and the type of processes that might be used in the future. The processes are (1) solvent refining, (2) the chemical removal of pyrite, and (3) the chemical removal of organic sulfur.

In the solvent-refining process coal is dissolved in an aromatic organic solvent under an atmosphere of hydrogen at moderate pressure. The solution is filtered to remove insoluble materials which consist of the ash, the pyritic sulfur, and a small amount of sulfur-containing, insoluble organic material. The resulting liquid is fractionally distilled to recover most of the solvent. In addition, small quantities of hydrocarbon gases and light liquids are produced. A heavy organic material called *solvent-refined coal* is left behind. This material contains less than 1% sulfur and has a heating value of about 16,000 Btu per pound regardless of the quality of the original coal used. The solvent-refining process removes all of the inorganic sulfur and 60—70% of the organic sulfur present in the original coal. A pilot plant is in operation to study the commercial feasibility of the process.

Most attempts to remove pyritic sulfur from coal have involved physical separations. The properties of iron pyrite have discouraged most attempts at chemical removal. For example, the pyrites are insoluble in most known liquids, including strong acids. Pyrites can be oxidized to water-soluble sulfates by strong oxidizing agents such as concentrated nitric acid or hydrogen peroxide, but in the process the coal matrix is also oxidized. Therefore, these reagents have not been considered practical.

The discovery of a new pyritic oxidizing agent has kindled new interest in the chemical removal of pyritic sulfur. The agent is very selective and oxidizes the pyrite without any significant effect on the coal matrix. The oxidizing agent is a water solution of iron (III) salts — $FeCl_3$ or $Fe_2(SO_4)_3$. These solutions oxidize the pyritic sulfur primarily to the free sulfur state. The free sulfur can then be removed from the coal by steam or vacuum vaporization. The resulting coal is essentially free of pyritic sulfur. The chemical steps in the process are represented as follows, where equation 7—6 represents a step in which the oxidizing agent is regenerated:

$$2Fe^{3+} + FeS_2 \cdot coal \longrightarrow 3Fe^{2+} + 2S \cdot coal \qquad \text{7—4}$$

$$2S \cdot coal \xrightarrow[\text{vaporization}]{} 2S + coal \qquad \text{7—5}$$

$$4Fe^{2+} + O_2 \text{ (from air)} \longrightarrow 4Fe^{3+} + 2O^{2-} \qquad \text{7—6}$$

This process is currently being evaluated under the sponsorship of the Environmental Protection Agency. An important advantage of the process is the use of an inexpensive oxidizing agent that can be regenerated easily. Table 7—7 contains data that illustrate the effectiveness of the process in removing pyritic sulfur from several selected coals.

The feasibility of extracting organic sulfur from coal without significantly changing the remainder of the coal matrix has been indicated in preliminary studies. Solvents have been found which selectively break down the coal matrix at the points occupied by organic sulfur compounds without disturbing the bonding in the major part of the coal matrix. Weak organic acids will remove 45—80% of the organic sulfur depending on process conditions and the type of coal used. Of these, nitrobenzene ($C_6H_5NO_2$) appears to be the most efficient solvent for the process. The process cost is very dependent on the degree of solvent loss, the ability to recycle the solvent, and the capacity of the solvent to dissolve large amounts of sulfur compounds.

The most immediately promising solution to the problem of sulfur in coal is the development of technology for removing sulfur oxides from exhaust gases. At least twenty processes for removing sulfur dioxide from

Table 7–7.   CHEMICAL REMOVAL OF PYRITIC SULFUR FROM COAL

| | Sulfur Content | | | |
Coal Source	Coal as Received (%)	Coal after Treatment (%)	Pyritic Sulfur Removed (%)	Total Sulfur Removed (%)
Randolph County, Illinois	3.81	2.19	98	43
Fulton County, Illinois	3.49	2.02	94	42
Greene County, Pennsylvania	1.77	0.77	84	57
Indiana County, Pennsylvania	4.30	1.13	88	74

Adapted with permission from "Desulfurization of Coal" by R. A. Meyers, et al., *Science,* vol. 177 (Sept. 29, 1972): Table 1, p. 1188. Copyright 1972 by the American Association for the Advancement of Science.

power plant exhausts are now being investigated. A number of these processes are being tested in full-scale units although none are yet commercially available.

No concerted efforts were made to remove sulfur dioxide from flue gases in the United States until 1966, when small-scale programs were first initiated. These programs were funded by both the government and private industry. Today the situation is much different. Federal expenditures dedicated to SO_2-emission control total more than $50 million per year. This amount is probably more than matched by money from private industrial sources.

Sulfur dioxide removal processes can be grouped into five general categories, each of which we will briefly discuss. The categories are dry injection, dry absorption, wet absorption, adsorption, and catalytic oxidation.

Dry injection is the simplest process and the most common material injected is limestone ($CaCO_3$). The limestone, in pulverized form, is injected into the firebox where it is calcined according to the following reaction.

$$CaCO_3 \xrightarrow{\text{high T}} CaO + CO_2 \qquad\qquad 7\text{--}7$$

The resulting calcium oxide (CaO) reacts at high temperature with SO_2 and excess oxygen in the firebox to form calcium sulfate ($CaSO_4$):

$$2CaO + 2SO_2 + O_2 \longrightarrow 2CaSO_4 \qquad\qquad 7\text{--}8$$

The solid calcium sulfate is removed along with fly ash by using mechanical and electrostatic precipitators.

The process appears to have an inherent limit to the amount of SO_2 that can be removed. Only a maximum of about 25% of the SO_2 present can be removed by a single injection of limestone. The process also has the problem of contributing to the difficulty of collecting fly ash. However, it requires the lowest capital outlay for equipment, and it has the lowest operating cost of any of the processes under development.

Therefore, it might be practical for use in smaller and older power plants where large capital investment cannot be tolerated. In addition, the process might be of use in plants using only marginally high-sulfur coal.

The operating principles of most absorption and adsorption processes are similar. The exhaust gases are brought into intimate contact with solids or liquids (the sorbers) contained in large chambers. The SO_2 in the exhaust gases reacts with the solids or liquids (absorption) or sticks to the surface of the solids (adsorption) and is removed. The cleaned exhaust gases are then released into the atmosphere. The solid or liquid sorbers are periodically replaced or regenerated and the collected sulfur compounds are often put to use.

In dry absorption processes the SO_2 is absorbed into a solid. The used absorbent is regenerated and gives up either concentrated SO_2 or other sulfur-containing compounds. Metal oxides as well as limestone have been used as absorbents. In one process of this type the absorbent is manganese dioxide (MnO_2) which reacts with absorbed SO_2 and produces manganese sulfate ($Mn_2(SO_4)_3$). The collected sulfur is ultimately recovered as commercially valuable ammonium sulfate ($(NH_4)_2SO_4$). A pilot plant using this process has reported removal efficiencies of nearly 90%, and future larger-scale tests are planned.

The major problems with dry absorption concern the absorbent itself. Absorbents must be very porous in order to have large surface areas and achieve as much reaction per cycle as possible. Researchers have found it quite difficult to develop suitable porosities and activities without seriously weakening the granule structure. Moreover, the absorbing activity is difficult to maintain because of the sintering or clumping together of granules caused by many regeneration cycles and because solid impurities in the exhaust gases tend to plug the pores of the absorbent.

At the present time wet absorption processes are thought to represent the best approach to exhaust gas cleanup. A wet lime/limestone process is especially attractive because the absorbents are cheap and the end products, mainly gypsum (hydrated $CaSO_4$), can be discarded with little adverse environmental effect. In one method, limestone is injected into the firebox just as it was in the dry injection process (see reactions 7–7 and 7–8). The resulting $CaSO_4$, unreacted CaO (lime) and SO_2, and fly ash are absorbed into water in a wet scrubber. In the water, further reaction between CaO and SO_2 takes place to form $CaSO_4$. The resulting solid $CaSO_4$ and fly ash are collected and discarded. The main problem with this technique is the tendency for calcium salts such as $CaSO_4$ to deposit on the surfaces of scrubbers and associated equipment.

Two other promising wet absorption methods are the Wellman-Lord and the sodium citrate processes. In the Wellman-Lord process a water solution of sodium sulfite (Na_2SO_3) is used as the absorbent. Dilute SO_2 from exhaust gases is concentrated by the following reactions:

$$SO_2 \text{ (dilute)} + Na_2SO_3 + H_2O \longrightarrow 2NaHSO_3 \qquad 7\text{–}9$$

Heat is applied to the concentrated solution of sodium bisulfite $(NaHSO_3)$, reaction 7—9 is reversed, and a concentrated SO_2 is released:

$$2NaHSO_3 \xrightarrow{\text{heat}} Na_2SO_3 + H_2O + SO_2 \text{(conc)} \qquad \textbf{7-10}$$

The collected SO_2 can be liquefied for further use or converted into marketable sulfur or sulfuric acid.

Exhaust gas to be cleaned by the sodium citrate process is first cooled and cleansed of particulate matter. It is then passed into an absorption tower and brought into contact with a solution containing citrate ions $(H_3 Cit^-)$. The following reactions take place:

$$SO_2 + H_2O \longrightarrow HSO_3^- \text{ (bisulfite ion)} + H^+ \qquad \textbf{7-11}$$

$$HSO_3^- + H_3 Cit^- \longrightarrow (HSO_3 \cdot H_3 Cit)^{2-} \qquad \textbf{7-12}$$

The solution now containing the bisulfite-citrate complex $[(HSO_3 \cdot H_3 Cit)^{2-}]$ is passed into a closed vessel and gaseous hydrogen sulfide $(H_2 S)$ is bubbled in. Sulfur precipitates out and is removed.

$$(HSO_3 \cdot H_3 Cit)^{2-} + H^+ + 2H_2S \longrightarrow 3S + H_3 Cit^- + 3H_2O \qquad \textbf{7-13}$$

The regenerated citrate solution is recirculated for further SO_2 removal and part of the precipitated sulfur is converted into the $H_2 S$ used in the process. As much as 99% of exhaust gas SO_2 has been removed by the use of this process.

An adsorption technique using activated carbon as the adsorbent has several potential advantages including low-temperature operation. In one such process, the Reinluft process, partially cleaned exhaust gas is passed upward through two adsorbent beds of coke. The SO_2 adsorbs to the coke and is removed from the exhaust gas. The used adsorbent is put into a desorbing unit where a reducing gas is used to liberate concentrated SO_2 and regenerate the coke. The SO_2 is collected and used as raw material for a sulfuric acid plant.

The catalytic oxidation (Cat-Ox) of SO_2 to yield sulfuric acid $(H_2 SO_4)$ is a promising alternative to the more conventional scrubbing procedures. In the Cat-Ox process, cleaned exhaust gas passes through a reactor where SO_2 is oxidized to sulfur trioxide (SO_3) with the aid of a vanadium pentoxide catalyst. The cooled exhaust gas, now containing SO_3, is next passed through a packed tower where $H_2 SO_4$ condenses out as a liquid. The exhaust gas is discharged into the atmosphere after going through a mist eliminator that removes tiny droplets of sulfuric acid. A 90% SO_2 removal and 80% resulting sulfuric acid concentration are possible with the process.

Despite the many processes potentially available for SO_2 removal, the necessary available technology is still in the prototype stage of development. No process has yet been sufficiently field-tested to declare it superior to other processes.

When SO_2 removal equipment becomes commercially available, another problem will present itself. During any given time, most utility operations can only remove a small part of their total capacity — that in excess of demand — off line for maintenance, repair, or modification. For example, Detroit Edison has reported that spare capacity would allow modification of only two plants a year. In 1973 Detroit Edison was operating twenty-three generating plants.

Coal-Mining Methods

Modern coal mining involves the systematic coordination of power equipment and personnel to extract coal by one of the following three methods: deep mining (underground), surface or strip mining, and auger mining. Deep mining is the most expensive method, followed, in order of decreasing costs, by strip and auger mining. A number of factors affect the productivity of a mine, or the percentage of available coal that can be recovered, and dictate the method of mining to be used as well. These factors include the thickness and structure of the coal bed, the depth of overburden, the characteristics of the strata above and below the coal, the dip (slope) of the coal bed, the presence of mine gas, and the presence of drainage water.

Deep mining is generally the rule when the overburden is more than 100 feet thick. The room-and-pillar technique predominates in deep mining. The coal face is usually undercut and drilled. Then explosives are used to break up the coal, and machines carry it to the surface by way of shafts and tunnels. The exposed roof of the mine is supported by coal pillars which are left in place to form networks of rooms and entries.

In areas where the overburden is 100 feet thick or less, strip and auger mining are considered to be practical. During a strip-mining operation, the relatively thin earth and rock overburden is removed to expose the coal bed and make it directly accessible to mining machines. The U.S. Geological Survey reports that between 1946 and 1970 the average thickness of overburden removed during strip coal mining increased from 32 to 55 feet. The maximum thickness removed increased from 70 feet in 1955 to 185 feet in 1970. The average weight-of-overburden-to-weight-of-coal ratio went from 6:1 in 1946 to 11:1 in 1970.

Auger mining can only be used under certain conditions since drill-like instruments are used to bore horizontally into coal seams. This method could be used, for example, on coal beds located near the surface of the side of a hill. The coal comes to the surface somewhat like wood chips do when a hole is drilled in a board.

The percentage of the available coal recovered from a mine is determined by economics, physical circumstances and the mining method used. For strip mines, recoveries of 80—90% are the rule as compared to approximately 50% for deep and auger mines. In deep-mining operations large losses are incurred as a result of leaving or only partially removing

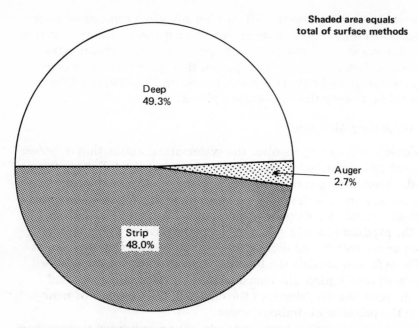

Shaded area equals
total of surface methods

Deep
49.3%

Auger
2.7%

Strip
48.0%

Figure 7—11.   Percent of U.S. Bituminous and Lignite Coal Produced by Various Mining
Methods, 1973

Data from *Coal Age* (Feb. 1974): 84. Reprinted by permission of McGraw-Hill, Inc.

the roof-supporting pillars. Figure 7—11 shows the percentage of total
U.S. bituminous and lignite coal obtained by each mining method in
1973. Note that the two surface methods accounted for over one half
of the total coal production. The amount obtained by surface mining has
steadily increased in recent years, as shown by Figure 7—12, which
contains statistics for the 10-year intervals between 1940 and 1970. The
main reason for the growth of strip mining in the U.S. is economic — the
method is cheaper than deep mining. In addition, underground mining is
considered to be more dangerous than surface mining.

All types of coal mining exert some undesirable effects on the
environment. For example, in some areas deep mines have collapsed and
caused surface ground to sink or have caught fire and produced air
pollution. However, no method of mining has created such a national
issue as strip mining. The environmental consequences usually associated
with strip mining have resulted in the introduction of congressional
legislation to ban it completely or at least carefully regulate it. We will
investigate the following three aspects of the problems associated with
strip mining:

1.   The unacceptable effects of strip mining.
2.   The consequences of a total ban on strip mining.
3.   Solutions to the problems other than a total ban.

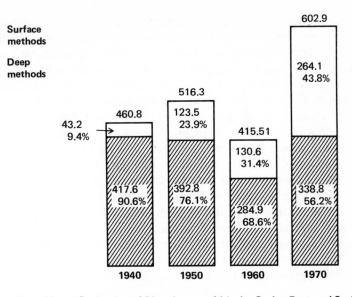

Figure 7–12.   United States Production of Bituminous and Lignite Coal at Deep and Surface
Mines, 1940–70 (Millions of tons and percent of total given)

Data from "Relationship of Energy and Fuel Shortages to the Nation's Internal Development,"
U.S. House of Representatives Subcommittee Hearings, Aug. 1972, p. 949.

The destruction of landscapes and soil associated with strip mining
is evident to everyone. For this reason strip mining has become a national
issue. The major environmental problems created by current strip mining
are soil erosion, water pollution, and the disruption of communities.

Erosion results when the overburden and mined sections of strip-
mined areas are left ungraded, improperly drained, and barren of
vegetation. Continual erosion results in the loss of topsoil and the
destruction of soil fertility. Sedimentation of water caused by soil erosion.
is a major water pollution problem below all strip-mined areas. The
effects of sedimentation include the destruction of fish habitats, clogged
culverts, etc., the narrowing of stream and river channels, and the
shortening of the useful lives of flood-control and water-storage dams.
The quantity of suspended sediment in the water below strip-mined
watersheds is more than 1000 times that in similar drainage basins where
no significant strip mining has occurred.

Another water pollution problem resulting from strip mining (and
other types of mining as well) is acid mine drainage, which is caused by
the presence of pyritic sulfur in coal deposits. Upon exposure to oxygen,
pyritic sulfur is oxidized to sulfuric acid, various sulfates, and iron oxides.
These substances enter surface water or groundwater and cause them to
become more acidic than normal. Water that is excessively acidic is
dangerous to aquatic life, corrosive, and less usable for recreational
activities.

Strip mining can also have a disrupting effect on local communities. Studies show that 9 of the 10 counties in West Virginia with the highest 1969—70 strip-mining production experienced significant population decreases. The average decrease was 17.6%. Similar population decreases have been documented for Kentucky and Ohio. After strip mining began in Belmont County, Ohio, there was an average 50% decrease in the assessed value of buildings as well as a decrease in land values.

Strip mines operate in different parts of the United States, but the extent of adverse effects varies. Figure 7—13 gives the percent of bituminous coal and lignite obtained in each state by strip mining during 1970. This figure seems to indicate that strip mining is a greater problem in the West than in the East. This, of course, is an invalid conclusion, for even though most coal in the West is obtained by strip mining, relatively little coal is produced. West Virginia, with only 10% strip mining, has much more severe problems because of the large quantity of coal mined. Thus, Figure 7—13 should be interpreted in conjunction with Figure 7—5, which gives the total coal production by state.

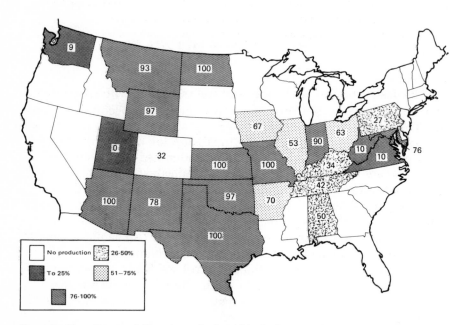

Figure 7—13.   Percent of Bituminous Coal and Lignite Produced by Strip Mining in the Conterminous United States, 1970

From "United States Coal Resources and Production," U.S. Dept. of the Interior, Bureau of Mines, June 1971, p. 22.

Because of the damaging effects previously discussed, a total ban on strip mining has been proposed by some. Opponents to a total ban point out that such action would not be without some major consequences, including the following:

1. Strip mining accounts for approximately 48% of the nation's total coal production. If this coal were eliminated by a total ban, the users (75% is used to generate electricity) would find it very difficult to find substitute fuels. Nuclear power has not been developed as fast as once anticipated and there is a shortage of domestic oil and gas reserves. If the electrical utilities converted to fuel oil, it would have to be imported. Assuming sufficient amounts of oil and delivery facilities were available, this purchase would cause an additional annual loss in the U.S. balance of payments.

2. It is not realistic to expect underground mines to be able to replace the production lost by a total ban on strip mining. While there are ample underground reserves, the production of 264 million tons of coal (the 1970 strip-mining production) would require 132 additional underground mines with a 2-million-ton annual capacity. This would require a capital investment of $3.2 to $3.7 billion. Full production from such mines would require 3—5 years to achieve and an additional 78,000 trained underground miners would be needed. Thus, the coal industry would have to virtually duplicate the present underground mine capacity with an enormous capital investment. At the same time, the coal industry would be required to write off as a loss any existing investments in strip-mining equipment and reserves.

3. A large part of the U.S. coal reserves can be mined only by surface methods. They lie under earth strata too shallow or unstable to safely support a roof. Thus, a total ban would effectively decrease the size of U.S. coal reserves. The switch to underground mining techniques would cause a reduction in reserves another way, since about half of the coal in an underground mine must be left in place to support the roof. In a strip mine the recovery approaches 100%.

4. To further complicate matters, most of the desirable low-sulfur coal reserves are of the type that cannot be mined by underground methods.

Strip-mining problems can be solved by action short of a total ban. The solutions involve the concept of reclamation which, for strip-mined land is simple in principle. After the coal is removed, the overburden must be returned to its original conformation — first the rock and then the topsoil. The aim of good reclamation is to reconstruct the land in order to restore or possibly increase its premining productivity. The uses for reclaimed land must be consistent with the nature of the soil, the topography of the land, the climate, and the uses of nearby land.

Laws now exist that make reclamation a mandatory part of any strip-mining project, and more are under consideration. These laws require that the final use of the land be planned before removing any coal. They also require that the mining process be followed as soon as is practical with action to shape the land, stabilize it against slides and erosion, and revegetate the surface.

Most of the reclamation laws deal with active strip mining and neglect a definite problem, the orphan banks. These banks were mined years ago when there were no legal requirements or public demand for reclamation. In many cases the involved mining companies are no longer in business and ownership or responsibility for the banks is impossible to fix. Many of the unsightly and eroded areas which now draw public criticism are orphan banks. Unfortunately, solutions to this problem have not yet been worked out. Table 7—8 indicates the proportion of land which must be reclaimed by law in those states having 50,000 or more acres needing reclamation.

Methods for Increasing Coal Use

A number of the problems associated with coal use can be alleviated by coal treatment processes now being perfected. In these processes, coal is converted into gaseous or liquid forms before use. The chemical treatments used also remove some of the undesirable impurities.

Table 7—8.   STATUS OF LAND DISTURBED BY COAL SURFACE MINING, JANUARY 1, 1974 (acres)

State	Reclamation Not Required by Law	Reclamation Required by Law
Pennsylvania	159,000	33,000
Kentucky	69,000	117,000
West Virginia	25,720	51,560
Montana	72,506	1,250
Illinois	49,748	20,891
Ohio	23,926	45,825
Alabama	57,878	118
Other states	164,109	67,437
Total	621,887	337,081

Data from U.S. Soil Conservation Service as quoted in "Surface Mining Control and Reclamation Act of 1974," U.S. House of Representatives Report No. 93-1072, p. 56.

Coal Gasification

Natural gas, as discussed in the previous chapter, is the most versatile and least polluting of the fossil fuels. However, increased demands have nearly exceeded the capacities of producers and distributors. A massive gap between natural gas supplies and demand is developing. Coal gasification, according to projections, will play an important role in diminishing the gap by providing an alternate supply of gaseous, clean-burning fuel in the near future.

The use of coal as a source of gaseous fuel has some very important advantages. Most important perhaps is the abundance of coal which was discussed earlier in this chapter. Another advantage is the fact that the coal supply is entirely domestic and is, therefore, a more reliable source than foreign imports of petroleum or liquefied natural gas. The use of domestic coal is also preferred from the standpoint of the critical balance-of-payments problem.

Coal gasification is the chemical transformation of solid coal into a gas. In this discussion we will include those process steps which result in the production of pipeline-quality gas. Pipeline-quality gas is composed essentially of methane (CH_4), is virtually free of sulfur, and contains no carbon monoxide or free hydrogen. Its heating value is about 1000 Btu/ft^3 as compared to a value of about 1050 Btu/ft^3 for natural gas. Pipeline-quality gas is often called *synthetic natural gas* (SNG), since it has properties very similar to those of natural gas.

Since the hydrogen content of coal (average of about 5%) is very low compared to that of methane (25%), coal gasification requires the addition of hydrogen to coal. During this process the coal matrix is hydrogenated, and in addition any sulfur, nitrogen, or oxygen constituents of the coal are converted respectively to hydrogen sulfide (H_2S), ammonia (NH_3), and water. The overall reaction requires several steps, however, and cannot be accomplished by a single direct hydrogenation.

A number of different coal gasification processes are being developed. However, they all use the same basic set of reactions, although not always in the same sequence. In a typical process the coal is first ground to a powder. It is then pretreated with air or oxygen to minimize its caking tendencies. This is necessary because heating causes some coals to swell and plug the reactor (gasifier). The basic device used in the process is the gasifier which is operated at pressures ranging from 20 to more than 70 atmospheres and temperatures as high as 1500°C. When the coal enters the gasifier, it is heated to drive off volatile constituents and to produce a gas rich in H_2 and CH_4. A solid residue called *char* is left behind after this devolitization. The reaction is represented by the following equation where the last C is char.

$$\text{Coal} \xrightarrow{600°C} H_2 + CH_4 + C \qquad \qquad \textbf{7–14}$$

The char is then reacted with superheated steam to form a gaseous product called *synthesis gas*. This product contains variable amounts of CO, H_2, and CH_4, which are considered to be valuable components, and CO_2 and sulfur compounds, which are considered to be undesirable impurities. The reactions that take place during synthesis gas formation are given below. This process is also known as the *hydrogasification step*.

$$C + H_2O \longrightarrow CO + H_2 \qquad \qquad 7-15$$

$$C + 2H_2 \longrightarrow CH_4 \qquad \qquad 7-16$$

$$CO + H_2O \longrightarrow CO_2 + H_2 \qquad \qquad 7-17$$

The resulting synthesis gas contains 40—65% CH_4. The synthesis gas is next passed over a *shift catalyst*, if necessary, where part of the CO reacts as follows:

$$CO + H_2O \xrightarrow{\text{catalyst}} CO_2 + H_2 \qquad \qquad 7-18$$

The objective of this catalytic shift conversion is to adjust the H_2/CO ratio to a value of 3:1 which is required for the next step, *catalytic methanation*. The reaction is:

$$CO + 3H_2 \xrightarrow{\text{catalyst}} H_2O + CH_4 \qquad \qquad 7-19$$

The synthesis gas must be carefully purified before the methanation step. Any CO_2, H_2S, organic sulfides, or water vapor present in the synthesis gas deactivate the methanation catalyst. After methanation the product (SNG) is essentially free of CO and sulfur compounds and has a heating value only about 50 Btu per ft^3 less than natural gas. Figure 7—14 contains a summary of the reactions involved in obtaining SNG from coal.

There are currently five major processes under development for coal gasification. They are listed in Table 7—9 along with the developing agency and pilot plant locations.

Only one of the five, the Lurgi process, is in commercial operation. It is used in Europe to produce a lower Btu gas than is required in the United States. Commercial-sized facilities based on the Lurgi process are anticipated in this country by 1976. The planned operations will use an extra methanation step to raise the heat content of the resultant gas to an acceptable level. Two such facilities are scheduled to be built in northwestern New Mexico by the Transwestern Coal Gasification Company and the El Paso Natural Gas Company. The plants are scheduled to begin gas production in 1976 and 1977, respectively.

All of the U.S. processes have been proven feasible on a laboratory scale. The various pilot plant programs are used to solve the mechanical and chemical engineering problems inherent in scaling a laboratory reaction up to commercial size. The production capacities of pilot plants

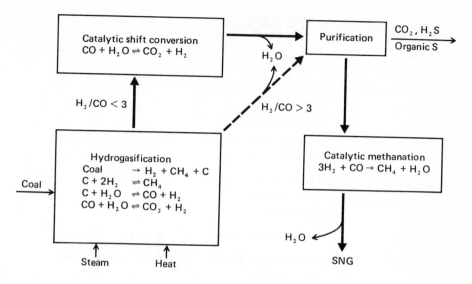

Figure 7—14.   A Generalized Flow Chart for Coal Gasification

Redrawn with permission from "Gasification: A Rediscovered Source of Clean Fuel," by Thomas H. Maugh, II, in *Science*, vol. 178 (Oct. 6, 1972): Fig. 1, p. 45. Copyright 1972 by the American Association for the Advancement of Science.

are in the range of 1—2 million ft^3 per day. The next step, a demonstration plant, will have a capacity of 60—80 million ft^3 per day, and a commercial plant will produce about 250 million ft^3 per day. Demonstration plants for the four U.S. processes are anticipated by 1976 with the design of full-sized plants possibly beginning in 1977. The full-sized plants could begin gas production by 1981. It is quite likely that after demonstration plants are constructed and evaluated, no single process will be found to be best for all situations in all parts of the country.

The price of gas produced by any of the gasification processes will be sensitive to coal costs. An increase of $1.00 per ton of coal will raise the price of produced gas by about $0.07 per 1000 ft^3. The final cost of synthetic natural gas has been estimated to be in the range of $0.65—$1.00 per 1000 ft^3.

A suitable site for a commercial gasification plant with a 250-million-ft^3-per-day capacity must include a nearby bituminous coal supply capable of providing 6 million tons per year or an equivalent amount of lower rank coal. In addition, a substantial supply of process and cooling water is essential. A quick calculation shows that this size of installation would require at least 120 million tons of uncommitted coal reserves to provide for a minimum 20-year life of the plant. Most uncommitted coal reserves are located in the western United States away from the energy-hungry population centers. Thus, the utilization of the full potential of coal gasification processes might depend upon the construction of pipelines or other distribution systems.

Table 7–9.   COAL GASIFICATION PROCESSES

Process Name	Process Developer	Pilot Plant Location
Hygas process	American Gas Association and Institute of Gas Technology, Chicago, Illinois	Chicago, Illinois
CO_2 Acceptor process	Consolidation Coal Company Pittsburgh, Pennsylvania	Rapid City, South Dakota
Bi-Gas process	Bituminous Coal Research Inc., Monroeville, Pennsylvania	Homer City, Pennsylvania
Synthane process	U.S. Bureau of Mines	Bruceton, Pennsylvania
Lurgi process	Lurgi Mineraloel Technik GmbH, West Germany	Commercial operation

From "Clean Energy from Coal — A National Priority," U.S. Dept. of the Interior, 1973 and adapted with permission from "Gasification: A Rediscovered Source of Clean Fuel," by T. H. Maugh, II, *Science*, vol. 178 (Oct. 6, 1972): 44–45. Copyright 1972 by the American Association for the Advancement of Science.

The production of gas from coal is not as new an idea as might appear from our discussion. Prior to World War II, *producer gas* was made from coal and used in some parts of the U.S. for industrial heating. It was made by passing air and steam through a thick bed of hot coal, and it consisted primarily of a mixture of N_2, CO, and H_2. Consequently, it had a low heating value — only one fifth that of natural gas — and today would not be interchangeable with natural gas in household furnaces or appliances. However, low-Btu gas of this type is once again attracting attention because of its possible use as a gas turbine fuel. The possibility of generating electricity by a combined gas turbine/ steam turbine cycle is providing the incentive.

Low-Btu gas for use in such electrical power generation might be produced by underground coal gasification. In October of 1972 the U.S. Bureau of Mines began underground gasification experiments. Holes were drilled into a 30-foot thick coal bed located 400 feet below the surface. The coal was then ignited by inserting a gas burner down a central bore-hole, also used to supply air from the surface. The useful applications of the technique are being evaluated.

Coal Liquefaction

It has been demonstrated that coal can be converted into liquid hydrocarbons that are suitable for producing gasoline and other products normally obtained from petroleum. The development of the first process of this type (the Bergius process) took place in Germany during World War I. This was followed in 1933 by the development of the Fischer-Tropsch process. During the Second World War, Germany used the Fischer-Tropsch method to produce 100 thousand barrels of liquid fuels per day. The first reaction of the process occurs when coke and steam are brought together at high temperatures:

$$C + H_2O \xrightarrow{high\ T} CO + H_2 \qquad\qquad 7\text{--}20$$

The resulting mixture of CO and H_2 is called *water gas* and can be used to produce a variety of hydrocarbons by treatment with various metal catalysts under proper conditions. Typical reactions are:

$$6CO + 12H_2 \xrightarrow{catalyst} C_6H_{12} + 6H_2O \qquad\qquad 7\text{--}21$$

$$7CO + 15H_2 \xrightarrow{catalyst} C_7H_{14} + 7H_2O \qquad\qquad 7\text{--}22$$

Neither the Bergius nor the Fischer-Tropsch process is economically feasible when petroleum supplies are plentiful. As a result, neither process has been attractive to U.S. industries. However, both processes represent potential hydrocarbon sources in case of necessity (such as wartime).

The development of modern, more economical processes for coal liquefaction is currently being supported by the U.S. Department of the Interior. However, the priority of the work is lower than that of coal gasification research and the development schedules are therefore less definite.

The U.S. process that has reached the most advanced stage of development (pilot plant demonstration) is the COED process — char-oil-energy-development process. In the COED process coal is crushed, dried, and then successively heated to higher and higher temperatures in a series of reactors. Typically, four stages (reactors) operating at 1022°, 1472°, 1742°, and 2642°C (600°, 850°, 1000°, and 1500°F) are used; in each reactor a part of the volatile matter of the coal is released. The number of stages used and the operating temperatures depend upon the tendency of the coal to agglomerate. Heat for the process is generated by burning char in the fourth stage and then using the resulting hot gases and hot char to heat the other reactors. The volatile products released from the coal pass into a product recovery system where oils are removed and gases are cooled. The recovered oil is filtered to remove solids and reacted with hydrogen under pressure to remove any sulfur, nitrogen, and oxygen that might be present.

A second process that is now in the pilot plant stage of development is the solvent-refined coal process which was discussed earlier in this chapter as a method for removing sulfur from coal. The product of the process, solvent-refined coal, melts at 350°C and therefore can be handled as a liquid.

A third process under active development utilizes direct coal hydrogenation. Several versions of the process are used and all are still confined to small-scale equipment. It has been demonstrated that finely ground coal can be hydrogenated in the solid state at elevated temperatures. Conversions of 70—80% of the coal have been obtained. The products contain 45—55% heavy liquids and 20—30% gases. At the present time the reactions are carried out in 1/8-inch diameter reactors that are 3 feet long. The coal is in contact with hydrogen inside the reactor for only one second or less — the coal is dropped through the vertical reactor. Unfortunately, the liquids produced in the process would, without further processing, be unsuitable as feed for a conventional petroleum refinery.

Suggested Readings

"Coal Gasification: How Best, How Soon," *Chemical and Engineering News*, November 12, 1973, pp. 18—19.

Department of the Interior, "Clean Energy from Coal — A National Priority," Government Printing Office, Washington, D.C., 1973.

————, "United States Coal Resources and Production," Government Printing Office, Washington, D.C., June 1971.

Dunham, J. T., C. Rampacek, and T. A. Henrie, "High-Sulfur Coal for Generating Electricity," *Science,* April 19, 1974, pp. 346—351.

Hill, G., "Some Aspects of Coal Research," *Chemical Technology,* May 1972, pp. 292—297.

Lessing, L. P., "Capturing Clean Gas and Oil from Coal," *Fortune,* Nov. 1973, pp. 129—131, 210, 214, 216, 218.

————, "Coal," *Scientific American,* July 1955.

Levene, H. D., "Gasification or Liquefaction: Where We Stand," *Coal Mining and Processing,* Jan. 1974, pp. 43—48.

Meyer, R. A., et al., "Desulfurization of Coal," *Science,* Sept. 29, 1972, pp. 1187—1188.

Mills, G. A., "Gas from Coal: Fuel of the Future," *Environmental Science and Technology,* Dec. 1971, pp. 1178—1182.

Nephew, E. A., "The Challenge and Promise of Coal," *Technology Review,* Dec. 1973, p. 21.

Office of Coal Research, Department of the Interior, "Coal Technology: Key to Clean Energy," Government Printing Office, Washington, D.C., 1974.

Osborn, E. F., "Coal and the Present Energy Situation," *Science,* Feb. 8, 1974, pp. 477—481.

Perry, H., "The Gasification of Coal," *Scientific American,* March 1974, pp. 19—25.

Slack, A. V., "Removing SO_2 from Stack Gases," *Environmental Science and Technology,* Feb. 1973, pp. 110—119.

"SO_2 Removal Technology Enters Growth Phase," *Environmental Science and Technology,* Aug. 1972, pp. 688—691.

Squires, A. M., "Clean Fuels from Coal Gasification," *Science*, April 19, 1974, pp. 340–346.
————, "Clean Power from Coal," *Science*, Aug. 28, 1970, pp. 821–828.
Walsh, J., "Problems of Expanding Coal Production," *Science*, April 19, 1974, pp. 336–339.

Chapter 8
Nuclear Energy

The nuclear age was born on December 2, 1942, in a squash court at the University of Chicago. On that day a group of scientists, led by Enrico Fermi, successfully operated the first nuclear reactor. Three years later, on the morning of July 16, 1945, nuclear energy was again released, this time in an uncontrolled, spectacularly violent manner: the first atomic bomb was detonated in the isolated deserts of southern New Mexico. A similar device was detonated about one month later (August 6, 1945) when the United States became the first nation to use a nuclear weapon in combat. Approximately 70,000 inhabitants of Hiroshima, Japan, were killed in the explosion and an equal number were seriously injured. The Second World War was brought to an end a short time later when another bomb was dropped on Nagasaki, Japan.

Since that time, scientists have been trying to apply nuclear energy to uses other than weaponry. These efforts can, according to some experts, ultimately lead to an unlimited source of controlled energy. President Nixon, in an April 18, 1973, message to Congress, stated that the major alternative to fossil fuels for the remainder of this century is nuclear energy. There are those who strongly disagree with this opinion on the basis of known and potential problems associated with nuclear energy use. Of course, advocates of nuclear energy use agree with the opinion and present arguments that tend to minimize the problems. In this chapter we will discuss some aspects of current and future nuclear energy use and the related problems.

Origin of Nuclear Energy

In 1896 Henri Becquerel, a French physicist, discovered that uranium compounds spontaneously emitted various rays that were capable of exposing photographic plates previously wrapped in light-proof paper. Substances with this property are said to be radioactive. The property of radioactivity has been found to be associated with instabilities of atomic nuclei. Some sort of rearrangement of nuclear particles or other energy transformations occur in high-energy (unstable) nuclei and energy is released, commonly in the form of alpha

(α), beta (β), or gamma (γ) radiation. Alpha rays consist of a stream of particles that are identical to the nuclei of helium atoms. Thus, each alpha particle contains 2 protons (p) and 2 neutrons (n) and carries a charge of +2. Beta rays are also composed of particles; the particles have been identified as electrons. Gamma rays are a form of electromagnetic radiation similar to visible light but having much higher frequencies and energies. Figure 8—1 contains a summary of these ideas.

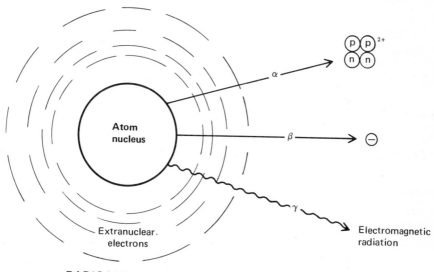

RADIOACTIVE ATOM

Figure 8—1.   Common Emissions from Radioactive Nuclei

Equivalence of Energy and Matter

An astute observation was made in 1903 by Ernest Rutherford and Frederick Soddy. They concluded that the nuclei of all atoms, not just the radioactive ones, must contain large quantities of energy. However, the form it took and the means for releasing it were unknown. In 1905, Albert Einstein, who was then working as a patent clerk, provided a partial answer in a part of his theory of relativity. He theorized that matter and energy were equivalent and interconvertible, as expressed by his now-famous equation:

$$E = mc^2 \qquad \text{8-1}$$

According to this equation, a small amount of matter of mass m would, upon conversion, yield a huge amount of energy E. The magnitude of the amount of energy liberated is indicated by the multiplier, c^2, which is the square of the velocity of light (c is approximately equal to 670 million miles per hour). The data of Table 8—1 show the amount of energy obtainable from the conversion of various amounts of matter.

Table 8—1.   AMOUNT OF ENERGY OBTAINABLE FROM COMPLETE CONVERSION OF MATTER TO ENERGY

Mass Converted	Btu	Energy Released Kilowatt hours	Calories
One electron (9.0×10^{-28} g)	7.7×10^{-17}	2.3×10^{-20}	1.9×10^{-14}
One proton (1.7×10^{-24} g)	1.4×10^{-13}	4.1×10^{-17}	3.5×10^{-11}
One carbon atom (2.0×10^{-23} g)	1.7×10^{-12}	5.0×10^{-16}	4.3×10^{-10}
One uranium atom (4.0×10^{-22} g)	3.4×10^{-11}	1.0×10^{-14}	8.6×10^{-9}
A virus (2.0×10^{-13} g)	1.7×10^{-2}	5.0×10^{-6}	4.3
One gram	8.5×10^{10}	2.5×10^{7}	2.1×10^{13}
8.1×10^{5} grams	**Total United States energy consumption, 1970**		
	6.9×10^{16}	2.0×10^{13}	1.7×10^{19}
1.7×10^{8} grams	**Earth's daily receipt of solar energy**		
	1.4×10^{19}	4.2×10^{15}	3.5×10^{21}
3.3×10^{18} grams	**Total daily output of energy by the sun**		
	2.8×10^{29}	8.2×10^{25}	7.1×10^{31}
The earth (6.0×10^{27} g)	5.1×10^{38}	1.5×10^{35}	1.3×10^{41}
The sun (2.0×10^{33} g)	1.7×10^{44}	5.0×10^{40}	4.3×10^{46}
Our galaxy (3.0×10^{44} g)	2.6×10^{55}	7.5×10^{51}	6.4×10^{57}

Although Einstein suggested that a possible verification of equation 8—1 might be found in the study of radioactivity, the idea lay dormant for a number of years. The realization that nuclear energy could be produced came early in 1939 when the work of Otto Hahn, Lisa Meitner, and Fritz Strassmann culminated in the discovery of nuclear fission. Notice that forty-three years elapsed between the first discovery of nuclear instability (Becquerel's 1896 discovery of radio-activity) and the proof of the feasibility of releasing nuclear energy. This is an important point to keep in mind when considering the development of new energy sources.

Nuclear Fission

The nuclear fission process discovered by Hahn, Meitner, and Strassmann represents an energy source of massive proportions. During nuclear fission large nuclei break into smaller fragments which have a total mass that is less than that of the original nuclei. The lost mass appears in the form of energy in accordance with equation 8—1. Nuclear fission takes place spontaneously in all nuclei containing a total number of protons and neutrons greater than 230. (The total number of protons and neutrons in the nucleus of an atom is called the *mass number.*) Spontaneous fission takes place much too slowly to constitute a practical energy source. However, the nuclei of a few heavy elements can be induced to undergo nuclear fission by bombarding them with other small particles. A specific isotope of uranium, $^{235}_{92}U$ or uranium-235, provides an example of this behavior. In the notation $^{235}_{92}U$, the superscript (235, in this case) is the mass number of the nucleus and the subscript (92) is the atomic number, or number of protons in the nucleus.

When a uranium-235 nucleus is hit by a relatively slow-moving, low-energy neutron (a so-called thermal neutron), the neutron is absorbed into the nucleus which becomes very unstable as a result. The unstable nucleus then breaks up into some combination of fragments such as those shown below. The mass difference between the particles on the left side of reaction 8—2 (a uranium-235 nucleus and a neutron)

$$^{235}_{92}U + ^{1}_{0}n \longrightarrow \text{unstable nucleus}$$

$$^{102}_{42}Mo + ^{132}_{50}La + 2\,^{1}_{0}n + \text{energy} \quad \textbf{8—2}$$

$$^{135}_{53}I + ^{97}_{39}Y + 4\,^{1}_{0}n + \text{energy} \quad \textbf{8—3}$$

$$^{139}_{56}Ba + ^{94}_{36}Kr + 3\,^{1}_{0}n + \text{energy} \quad \textbf{8—4}$$

$$^{139}_{54}Xe + ^{95}_{38}Sr + 2\,^{1}_{0}n + \text{energy} \quad \textbf{8—5}$$

and the particles on the right (a molybdenum-102 nucleus, a tin-132 nucleus and 2 neutrons) is equivalent to about 3.22×10^{-4} ergs or 3.06×10^{-14} Btu. Most of the energy liberated in this and other fission reactions appears as kinetic energy of the fission fragments and neutrons (89%). The remaining 11% is carried off as radioactive emissions. Most of the kinetic energy of the fission fragments is quickly transferred to the surroundings and appears as heat, a form that can be used directly or easily changed into other useful forms such as electricity.

The amount of energy available from a uranium fission process can be illustrated easily. A 235-gram sample (about ½ pound) of uranium-235 contains 6×10^{23} fissionable nuclei and has a volume represented by a cube slightly less than one inch on a side. If we assume all of the nuclei in this sample undergo fission according to reaction 8—2, the total energy released is calculated as follows:

$$(6 \times 10^{23} \text{ nuclei})(3.06 \times 10^{-14} \text{ Btu/nucleus}) = 1.8 \times 10^{10} \text{ Btu}$$

This amount of energy could provide 1000 six-room homes with a full year's supply of heat and electricity.

Table 8—2 provides a comparison of the relative energy content of the traditional fossil fuels and fissionable uranium. These data clearly illustrate how the widespread use of nuclear fission could conserve fossil fuels.

Table 8—2.   ENERGY EQUIVALENCE OF VARIOUS SOURCES

Energy Source	Unit	Energy/Unit (thousands of Btu)	Amount Equivalent to One Barrel of Petroleum
Petroleum	Barrel	5,800	1 barrel
Natural gas	Cubic foot	1.032	5620 cubic feet
Coal	Ton	26,000	0.223 tons
Uranium	Gram	76,600	0.076 grams
Uranium	Pound	36,600,000	0.000158 pounds

Adapted from "An Analysis of the Economic and Security Aspects of the Trans-Alaska Pipeline," vol. 3, Supplement Energy and Policy Alternatives, U.S. Dept. of the Interior, Office of Economic Analysis, March 1972, Table A-1, p. A-2.

Nuclear Fusion

Curiosity about the nature of the apparently unlimited energy supply of the stars (especially the sun) led to an understanding of a second energy-producing process — nuclear fusion. Sir Arthur Eddington suggested in 1920 that the energy of the stars was a by-product of the conversion of hydrogen into helium. This was followed in 1929 by the concept of thermonuclear or high-temperature-induced nuclear reactions. According to this concept, the nuclei of lightweight elements will combine, or fuse, when heated to very high temperatures and produce heavier nuclei. Thus, the nuclear fusion process is somewhat analogous to the reverse of nuclear fission.

Fusion reactions liberate energy for the same reason as fission reactions — the products have less total mass than the reactants and the difference appears as energy. The fusion reactions of the sun, though not fully understood, illustrate the process. As Eddington proposed, hydrogen is converted into helium. The reactions occur at such high temperatures that all reactants and products are completely stripped of their extranuclear electrons, and it is really the charged nuclei that react. The overall fusion reaction of the sun takes place in steps:

1. Two hydrogen nuclei ($_1^1 H^+$ or $_1^1 p$) fuse to produce a deuteron ($_1^2 H^+$ or $_1^2 D^+$), a positron (e^+ or β^+) and a neutrino (ν^0). A *deuteron* is a particle consisting of one proton and one neutron, a *positron* is a positively charged electron, and a *neutrino* is a very small uncharged particle. This step is represented by the following reaction:

$$_1^1 H^+ + {_1^1} H^+ \longrightarrow {_1^2} D^+ + e^+ + \nu^0 \qquad\qquad 8\text{—}6$$

2. The deuterons of reaction 8—6 fuse with other hydrogen nuclei to form helium-3 nuclei and gamma rays:

$$^2_1D^+ + ^1_1H^+ \longrightarrow ^3_2He^{2+} + \gamma^0 \qquad 8\text{--}7$$

3. Pairs of helium-3 nuclei fuse to form ordinary helium-4 nuclei plus more hydrogen nuclei:

$$^3_2He^{2+} + ^3_2He^{2+} \longrightarrow ^4_2He^{2+} + ^1_1H^+ + ^1_1H^+ \qquad 8\text{--}8$$

The net overall fusion reaction of the sun, as represented by reaction 8–9,

$$4\,^1_1H^+ \longrightarrow ^4_2He^{2+} + 2e^+ + 2\nu^0 + 2\gamma^0 \qquad 8\text{--}9$$

shows that the sun is consuming hydrogen fuel and producing helium ashes. The mass difference between the right and left sides of reaction 8–9 accounts for the release of about 4×10^{-15} Btu of energy, or about 1×10^{-15} Btu per hydrogen nucleus reacted. According to Table 8–2, one gram of fissionable uranium produces 76.6 million Btu. One gram of hydrogen, undergoing fusion according to reaction 8–9, produces 600 million Btu, or about 8 times as much as the uranium. On this basis, 0.0085 grams of hydrogen is equivalent to a barrel of petroleum.

The fusion reactions of the sun take place in a central core where the temperature is estimated to be 15 million degrees Kelvin (also about 15 million °C or 27 million °F). About 2×10^{19} kilograms (22 quadrillion tons) of hydrogen is consumed each year. This process has gone on for an estimated 4.5 billion years and will continue for another 3 billion years, at which time the hydrogen in the central core (about 5% of the total) will have been used up. At this time the hydrogen will begin to react in a shell around the core, the sun will expand, and the earth will be faced with another energy crisis — too much energy.

Unlike the situation with fission reactions, no energy has yet been obtained from controlled fusion reactions carried out on the earth. Hydrogen bombs have been detonated but the reactions have obviously not been under control. In fact, a fission bomb must be detonated to achieve the high temperatures necessary to start fusion. The fusion reactions of a hydrogen bomb, which involve $^2_1D^+$, $^3_1T^+$ (tritium), and $^6_3Li^+$, are:

$$^3_1T^+ + ^2_1D^+ \longrightarrow ^4_2He^{2+} + ^1_0n \qquad 8\text{--}10$$

$$^6_3Li^{3+} + ^1_0n \longrightarrow ^4_2He^{2+} + ^3_1T^+ \qquad 8\text{--}11$$

$$^2_1D^+ + ^3_1T^+ \longrightarrow ^4_2He^{2+} + ^1_0n \qquad 8\text{--}12$$

Vast amounts of money and untold hours of work have been expended in attempts to harness fusion reactions for use as energy sources. These efforts will continue, for fusion energy has a number of distinct advantages over currently used energy sources. These advantages and the progress made toward obtaining fusion energy are discussed later in this chapter.

Controlled Generation of Nuclear Energy

The uses of nuclear energy are varied and range from the destructive weapons of war to healing techniques used in medicine. Bombs or other nuclear explosives, as mentioned before, represent an essentially uncontrolled release of nuclear energy and at this time are the only uses made of fusion reactions. However, the energy of fission reactions can be released in a controlled way and converted into useful forms such as electricity.

Chain Reactions

Earlier we saw that some nuclei, such as uranium-235, can be induced to undergo fission by neutron bombardment. Notice that in reactions 8—2 through 8—5 neutrons are produced by the fission of a uranium-235 nucleus. This creates the possibility for chain reactions. If only one of the neutrons produced each time reacts with another uranium-235 nucleus, a chain reaction results, and the fission process proceeds at a constant rate. When such conditions prevail, the process is said to be self-propagating, or *critical.* If more than one of the generated neutrons causes another reaction, the chain reaction of the critical situation becomes a branching chain, and a *supercritical* condition is created which leads to an explosion. Critical and supercritical reactions are depicted in Figure 8—2.

Critical Mass

If a single uranium-235 atom were isolated and caused to undergo fission, it is obvious that no further reactions could be caused by the generated neutrons. If two U-235 atoms were collected together, the chance for a fission-generated neutron to cause a reaction in the nucleus of the second atom is increased. As the number of atoms in the collection is increased, the chance for a generated neutron to induce fission in other nuclei also increases. It seems apparent that a sample could be collected which contained just enough atoms to insure the existence of a self-propagating or critical condition. The minimum amount of fissionable material that must be put together in one piece to create such a situation is called the *critical mass.* An amount larger than the critical mass would create a supercritical condition and an explosion. This was the principle used in the early atomic bombs. Subcritical masses of fissionable material were brought together at the time of detonation to form supercritical masses. The subcritical masses were forced together by conventional explosive charges as shown diagrammatically in Figure 8—3. The different configurations resulted in two different bomb shapes as shown in Figure 8—4. The spherical configuration was housed in the "fat man" bomb casing (Figure

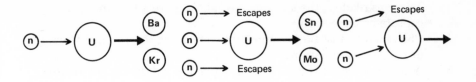

CRITICAL

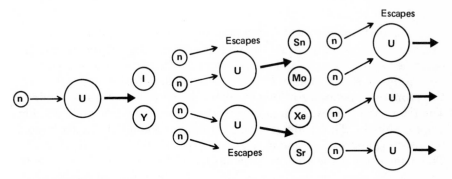

SUPERCRITICAL

Figure 8—2.   Nuclear Chain Reactions

From *Chemistry: A Science for Today* by Spencer L. Seager and H. Stephen Stoker, p. 250. Copyright © 1973 by Scott, Foresman and Company. Reprinted by permission of the publisher.

8—4 A); the tubular configuration gave rise to the "little boy" bomb (Figure 8—4 B). Figure 8—4 C is the casing of the device used in the Project Rulison experiment to test the feasibility of stimulating natural gas wells by nuclear explosives (see Chapter 6).

Nuclear Reactors

The controlled release of nuclear energy from fission reactions is accomplished by using a nuclear reactor. There are five essentials of such reactors:

1. A quantity of fissionable material sufficient to create a self-sustaining or critical nuclear reaction must be present.
2. A moderator is needed to slow the high-energy neutrons produced during fission. The slower neutrons have a higher probability of colliding with other fissionable nuclei and causing further fission reactions.
3. A means for controlling the rate of the resulting reaction is required. This is usually some substance with the ability to absorb

neutrons readily; the substance is said to have a large neutron capture cross section.

4. The reactor must be surrounded by shielding to prevent the escape of dangerous radiation.

5. A cooling system is necessary. Heat generated during the fission process must be removed to prevent the reactor components from being melted or otherwise damaged by extreme temperatures. The heat carried to the outside of the reactor is the energy that can be used directly or converted into other forms; the latter is the usual practice.

The first nuclear reactor was built under the direction of Enrico Fermi and successfully tested in 1942. It contained all of the essentials described above except for a cooling system. This reactor was designed only to prove the feasibility of controlling nuclear fission and was therefore not operated for long enough time periods to require cooling. The fissionable fuel was uranium metal and uranium oxide embedded in graphite blocks. The moderator was graphite and control was achieved by inserting appropriate metal rods into holes built into the reactor. The name often associated with the device, atomic pile, was quite appropriate, since the reactor was constructed by piling up fuel-containing graphite blocks and moderator graphite blocks in alternated layers. After 57 layers had been assembled, the reactor reaction became self-sustaining. The Fermi atomic pile is represented schematically in Figure 8–5.

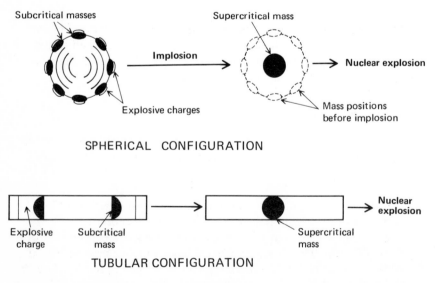

SPHERICAL CONFIGURATION

TUBULAR CONFIGURATION

Figure 8–3.   Configurations Used in Fission Bombs

Figure 8—4. Containment Cases for Nuclear Devices — (A) "Fat Man," (B) "Little Boy," (C) Rulison Device

From the Los Alamos Scientific Laboratory brochure, May 1971, p. 12.

Fission Reactors

The components of a commercial facility used to produce electrical energy from controlled fission reactions are shown schematically in Figure 8—6.

The coolant used in commercial power reactors may be either a gas or a liquid. Gases such as air, helium, and carbon dioxide are currently used. The liquids now in use include ordinary water, heavy water, organic compounds, and even liquid (molten) metals such as sodium and lithium. In some instances the coolant surrounding the fuel-containing core also acts as a moderator. The heat carried from the reactor may be transferred to a secondary fluid as depicted in Figure 8—6 or the hot coolant (or coolant vapor) can be used directly to drive the generator turbine.

The reactor core, the heart of the system, contains fuel elements which have been fabricated into forms such as rods, tubes, or plates. In most reactors the fuels within the fuel elements are solids, but in a few instances liquid fuels, such as water solutions of uranium salts, have been used.

Typically, reactors are equipped with two sets of control rods. One set is used for ordinary regulatory purposes to control the rate of the fission reactions and the energy output of the system. A second set,

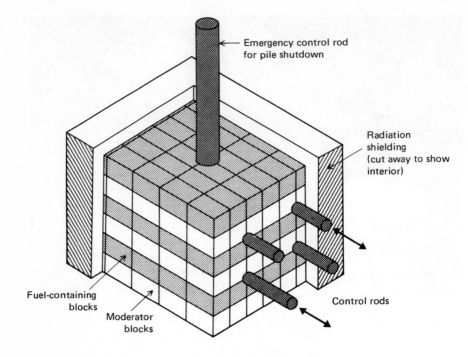

Figure 8—5.   The Fermi Atomic Pile, The World's First Nuclear Reactor

used only in the event of an emergency, can be employed to shut the
reactor down very rapidly.

A wide variety of different fission-power reactor systems have
been designed or proposed based on different fuels, moderators, and
coolants. An assessment of the United States' reactor development
program was conducted in 1964 in an attempt to determine the
developmental status and potential of each reactor concept then under
consideration. As a result, a number of systems were abandoned and an
emphasis was placed on the development of just a few reactor concepts.
Those chosen for continued development were reactors in which
ordinary water is used as a moderator and coolant — the light water
reactors (LWR) — and the high-temperature gas-cooled reactors
(HTGR). Light water reactors were chosen because their development
was already well advanced as a result of activities by the power-
generating industry. A number had been installed and proven reliable,
safe, and economical to operate.

Light Water Reactors

Light water reactors are classified into two categories, boiling
water reactors (BWR) and pressurized water reactors (PWR). A simplified illustration of a BWR power generation system is given in

Figure 8–7. Boiling water, forced through the reactor core, acts both as a moderator and coolant. The water is heated to a temperature of about 135°C (275°F), and steam is generated at a pressure of approximately 1000 pounds per square inch. The steam is used to drive the turbine of an electrical generator. Physically, boiling water reactors are large. The pressure vessel housing the reactor core of a 500-megawatt unit is 19 feet in diameter and 60 feet high. The complexity of the reactor is

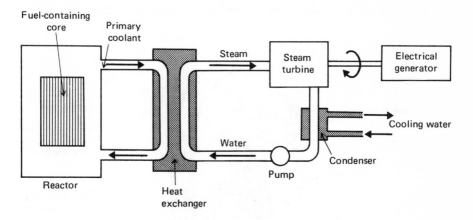

Figure 8–6.   Components of a Nuclear Electricity Power Plant

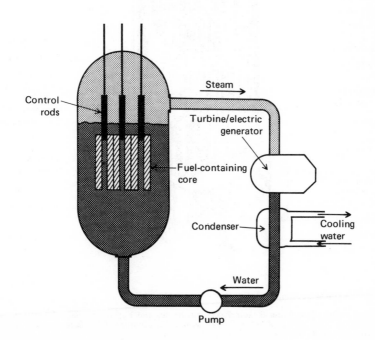

Figure 8–7.   Energy Production from a Boiling Water Reactor

illustrated by the detailed representation of the reactor pressure vessel shown in Figure 8—8.

A pressurized water reactor system is represented in Figure 8—9. Once again, water serves as both a coolant and moderator. The water is maintained at a pressure of about 2000 pounds per square inch, which allows it to be heated to 300°C (572°F) without boiling. The hot water is pumped to a heat exchanger where much of the heat is transferred to a secondary water-filled loop maintained at a lower pressure. Steam generated in the secondary loop is used to drive an electrical generator.

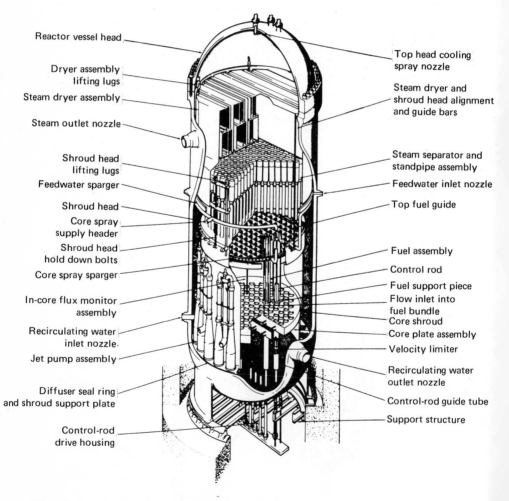

Figure 8—8.   Typical Boiling Water Reactor Pressure Vessel

From "United States Light-Water Reactors: Present Status and Future Prospects," by W. K. Davis, et al., in *Peaceful Uses of Atomic Energy,* vol. 2, Proceedings of the International Atomic Energy Agency, 1972, p. 28.

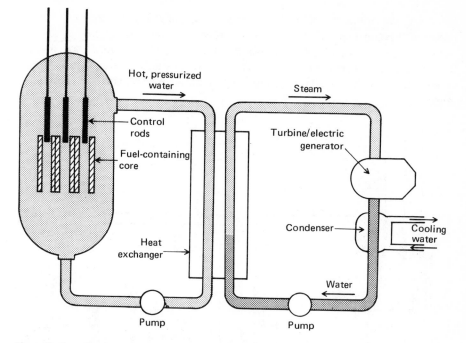

Figure 8–9. Energy Production from a Pressurized Water Reactor

High-Temperature Gas-Cooled Reactors

The high-temperature gas-cooled reactor represents an improvement over light water reactors. The HTGR efficiency is about 40%, which compares favorably with most modern coal-fired plants (39%) and is about 10% higher than that of light water reactors. The HTGR system has been under development in the United States since 1958. A 40-megawatt prototype installation (the Peach Bottom plant) has been in operation since 1967, and a 330-megawatt demonstration plant began operation in 1974. A total of 6 more HTGR plants are scheduled to be operating by 1982. Data from the Peach Bottom plant were used to determine the HTGR efficiency given above.

Graphite is used as a moderator and helium gas is the preferred coolant for HTGRs. Helium gas is used because it has good thermal transport properties and is chemically inert. The inertness of helium is important because of the potential for corrosion that exists at the high operating temperature of the reactor. The use of helium presents at least one problem, that of containment. Helium can escape through extremely small openings, so reactor vessels must be leakproof. This requirement increases the expense involved in constructing HTGR systems.

During operation of the HTGR helium gas at a pressure of 300—400 pounds per square inch is forced through the reactor core where it is heated to 750° C (1380° F). The hot gas is pumped to a heat exchanger where water in a secondary loop is boiled, as diagrammed in Figure 8—10. The resulting steam is used as before to produce electrical energy.

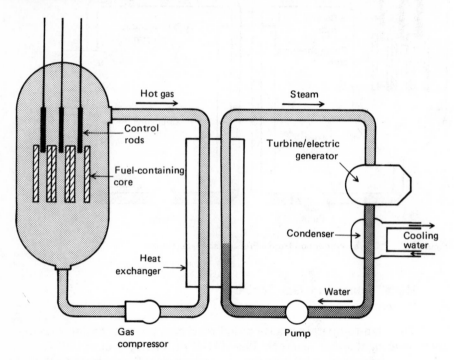

Figure 8—10.   Energy Production from a High-Temperature Gas-Cooled Reactor

The status of nuclear power generation in the U.S. from LWR and HTGR systems is indicated by the data of Table 8—3.

Breeder Reactors

United States Atomic Energy Commission studies during the early 1960s indicated that a power development program based almost entirely on LWR technology (as implied by Table 8—3) would cause nuclear fuel prices to soar. The increasing demand would make it necessary to extract the fuel from less desirable and more expensive sources. Figure 8—11 illustrates the problem. According to these estimates, the total reserves of fuel costing $10 or less per pound would be committed for use by about 1993, and those costing up to $15 a pound would be committed before the turn of the century. These projections led to the introduction of a high priority program to develop an advanced reactor for electric power production involving what is now

known as the "breeder" reactor concept. The breeder gets its name
from the fact that it will produce more nuclear fuel material than it
consumes. Breeder reactors have the capability of converting non-fission-
able substances into fissionable nuclear fuels.

The fissionable material used in light water reactors is
uranium-235. However, this isotope constitutes only 0.72% of naturally
occurring uranium; the other 99.27% is made up of nonfissionable
uranium-238. The actual fuel elements of light water reactors contain
uranium oxide (UO_2) which has been enriched (a process discussed
later) so that 2.5—3.5% of the uranium present is uranium-235. The
fission reactions of light water reactors are similar to those discussed
earlier (reactions 8—2 through 8—5). Some of the fission products have
high neutron capture cross sections (ability to capture neutrons) and
therefore act somewhat like control rods to slow the fission reactions.
As the concentrations of such products (nuclear fuel "poisons")
increase within a fuel element, the rate of the fission process decreases
until the process becomes very inefficient and could cease. For this

Table 8—3.   STATUS OF LWR AND HTGR POWER DEVELOPMENT IN THE UNITED
STATES, JANUARY 1, 1974

(includes units operable, under construction, or on order)

Year of Scheduled Operation	LWR		HTGR	
	Number of Units	Total Capacity (megawatts)	Number of Units	Total Capacity (megawatts)
Cumulative through 1974	57	39,080	2	370
Additional Units				
1975	11	10,555	0	0
1976	9	9,222	0	0
1977	8	7,884	0	0
1978	8	8,013	0	0
1979	17	17,248	0	0
1980	27	28,478	1	770
1981	26	29,259	2	1,910
1982	16	18,263	2	1,540
1983	6	6,577	1	1,140
1984	2	2,570	0	0
1985	2	2,450	0	0
1986	2	2,450	0	0
Total	191	182,049	8	5,730

Data from "The Nuclear Industry 1973," U.S. Atomic Energy Commission Publication WASH
1174—73, Government Printing Office, 1974, pp. 4—8.

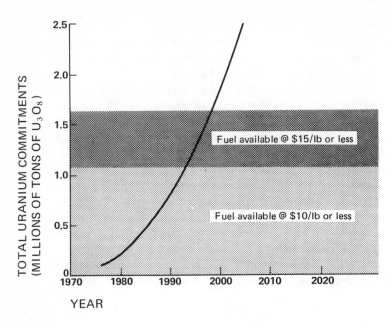

YEAR

Figure 8–11.   Nuclear Fuel Commitment to LWR and HTGR Energy Production

Modified from "Outlook for Uranium Production to Meet Future Nuclear-Fuel Needs in the United States of America," by R. L. Faulkner in *Peaceful Uses of Atomic Energy*, vol. 8, Proceedings of the International Atomic Energy Agency, 1972, Fig. 7, p. 32.

reason, fuel elements are periodically removed from the reactor and sent to chemical reprocessing plants where any remaining fissionable material is recovered and the fuel poisons are removed. The recovered fissionable materials are formed into new fuel elements and returned to the reactor.

The fissionable material recovered from used fuel elements during reprocessing is found to consist of unreacted uranium-235 and a totally new substance called plutonium (Pu). This substance does not occur naturally; it is a synthetic element. Plutonium is produced by a reaction between uranium-238 and neutrons generated by the fission of uranium-235. Thus, some nonfissionable uranium-238 is converted into a useful, fissionable nuclear fuel.

$$^{238}_{92}U + ^{1}_{0}n \longrightarrow ^{239}_{94}Pu + 2^{0}_{-1}\beta \qquad \qquad 8\text{–}13$$

Light water reactors are not classified as breeder reactors because the number of fissionable plutonium atoms generated is less than the number of uranium-235 atoms consumed during operation. However, the process increases the amount of energy that can be obtained from a uranium fuel mixture since part of the uranium-238 becomes fuel for the reactor.

High-temperature gas-cooled reactors come closer to being true breeders; they are capable of producing practically as much fuel as they consume. The fuel elements of an HTGR consist of a mixture of highly enriched (about 90% uranium-235) uranium compounds (UO_2 or UC_2) and thorium compounds (ThO_2 or ThC_2). The uranium-235 undergoes fission and some of the generated neutrons react with thorium-232 (the only naturally occurring thorium isotope). Uranium-233, a fissionable isotope, is produced.

$$^{232}_{90}Th + ^{1}_{0}n \longrightarrow ^{233}_{92}U + 2^{-0}_{-1}\beta \qquad\qquad 8\text{--}14$$

The use of thorium as a conventional (nonbreeder) reactor fuel conserves rather than replaces uranium since uranium is still needed to start the conversion of thorium into a fuel. However, the increased use of thorium in the future could exert a profound influence on uranium consumption because thorium is about 3 times more abundant than uranium.

True breeder reactors generate more nuclear fuel than they consume. The excess generated fuel can be extracted during reprocessing and used in other reactors. Breeder reactors are classified into two categories on the basis of the energies of the neutrons involved in the process. Fast breeders make use of neutrons with energies 10,000 to 100 million times greater than the slower moving thermal neutrons of the thermal breeders. In fast-breeder reactors, fissionable uranium-235 or plutonium-239 is used to convert uranium-238, the fertile material, into plutonium-239 (reaction 8—13). The lower energy neutrons of thermal breeders are used to convert thorium-232 into uranium-233 (reaction 8—14). It must also be kept in mind that breeder reactors produce usable energy as well as nuclear fuels.

Four main types of breeder reactors are being developed in the United States: liquid metal fast breeders (LMFB), gas-cooled fast breeders (GCFB), light water breeders (LWB) and molten salt breeders (MSB). The liquid metal fast-breeder concept has the highest developmental priority in both the United States and foreign countries. At least twenty-two LMFB projects were underway throughout the world in 1973. Three of these were in the United States.

The coolant used in most LMFB systems is molten sodium metal. This material has excellent heat transfer properties and its $895°C$ boiling point allows the system to be operated at low pressures. An added advantage of liquid sodium use is the ability of the coolant to cool by natural circulation in the event of an emergency shutdown. However, liquid sodium use also has some disadvantages. It becomes radioactive when it passes through the reactor core. This necessitates the use of a secondary sodium loop and an extra heat exchanger to transfer reactor heat to an electrical generator. Another disadvantage is the reactivity of sodium with water. Upon contact, these two materials react rapidly and produce hydrogen gas, which forms an explosive mixture with air. Thus, it is necessary to develop commercial steam generators that either eliminate the possibility of direct leaks of water into sodium or minimize the effects of such leaks. A schematic representation of an LMFB system is given in Figure 8—12.

Gas-cooled fast breeders (GCFB) offer an alternative to the LMFB. They have the potential for breeding ratios of 1.4, or greater with stainless-steel clad oxide fuel, and fuel doubling times of about 10 years. An intermediate heat exchange loop between the core and steam system is not required. The GCFB program also has the advantage of being a natural continuation of the ongoing HTGR program. The general design concepts of a GCFB system are similar to those of an HTGR system (see Figure 8—10). Helium gas, used as a coolant, has the advantages of being chemically inert and resistant to induced radioactivity. In addition,

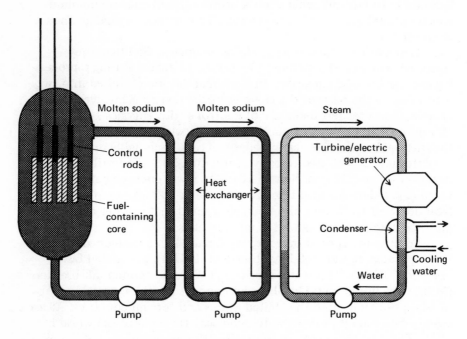

Figure 8—12.   A Liquid Metal Fast-Breeder Reactor System

it allows the reactor core to be observed during refueling operations — not easily done with liquid metal coolants. Both the reactor and heat exchanger are contained in a prestressed concrete pressure vessel about 84 feet in diameter and 70 feet high. The reactor core itself stands only 3 feet high. The hot helium delivered to the heat exchanger is under a pressure of 1250 to 1750 pounds per square inch. The main disadvantage of a gas coolant is the possibility of suddenly losing coolant pressure and thus cooling capacity for the reactor. Attempts to prevent such events have led to problems in designing the containment system.

The light water breeder reactor is essentially a pressurized water reactor (see Figure 8—9) with a modified core design. These reactors produce just enough fuel to satisfy their own needs and so are marginally effective as breeders. The advantage of the system is that the existing PWR technology can be used with little further development.

The molten salt breeder reactor was developed in an attempt to perfect a system that could operate for long periods of time at high temperatures with good efficiency. On-site fuel reprocessing plants are used to help reduce fuel costs. During operation, a high-temperature fuel mixture is circulated in liquid form through the system. The fuel consists of a mixture of uranium and thorium compounds dissolved in molten lithium fluoride (LiF) and beryllium fluoride (BeF_2) carrier salts. The advantages of the system include a built-in chemical reprocessing plant, the elimination of metal fuel containers, the elimination of refabricating the fuel into some geometric shape, and the ability to remove rapidly any undesirable fission products. The primary disadvantage is the necessity of circulating highly radioactive material throughout the system. A small (less than 10-megawatt) experimental MSB reactor was operated for nearly 5 years in the United States — until December of 1969. These experiments revealed some potential difficulties with materials used in the primary circulation system. Metallurgical research is underway in an attempt to solve the problems.

Table 8—4 contains data related to the performance characteristics for three of the breeder reactor systems discussed. These characteristics will continue to be evaluated as further development and testing proceed. The United States is committed to build a 400-megawatt Fast Flux Text Facility (FFTF) in Richland, Washington. This project, scheduled for completion in 1977, will contain the largest test reactor

Table 8—4. TYPICAL PERFORMANCE PARAMETERS FOR 1000-MEGAWATT BREEDER REACTORS

Reactor Type	Fissionable Fuel Present (kg)	Specific Power Output (megawatts/kg)	Fertile Material Present (kg)	Fuel Doubling Time (years)
Liquid Metal Fast Breeder	3900	0.26	70,000	15
Gas-Cooled Fast Breeder	5000	0.20	120,000	10
Molten Salt Fast Breeder	1500	0.67	70,000	21

Adapted with permission from "Energy from Breeder Reactors" by Floyd L. Cutter, Jr., and William O. Harms in *Physics Today* (May 1972): Table 5, p. 37. Reprinted with permission from the American Institute of Physics, Inc.

facility in the world designed specifically to test nuclear fuel, assemblies, and materials up to the point of failure. Further information will be obtained from the first large-scale demonstration plant using an LMFB reactor. This facility, expected to be in operation sometime between 1978 and 1980, will have an electrical power output of 350–400 megawatts. It will be built on a site on the Clinch River in east Tennessee at an estimated cost of $700 billion. The plant will be operated by the Tennessee Valley Authority under the direction of the Commonwealth Edison Company and its output fed into the TVA distribution system.

Production and Use of Nuclear Energy

The nuclear energy industry began on December 2, 1957, with the operation of the first commercial (fission) reactor at Shippingport, Pennsylvania. The industry has grown steadily since then, as shown by the nuclear power plant licensing trends given in Figure 8–13. At the end of 1973, 42 plants were licensed to operate, and all but two were

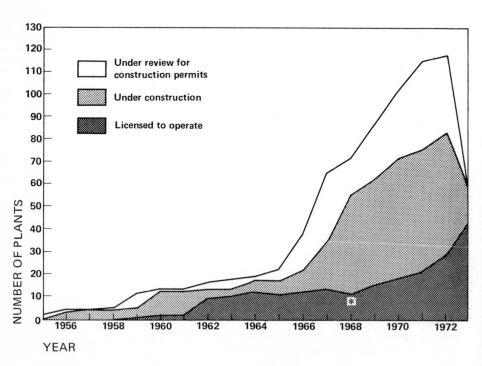

* Fluctuation in numbers of operating reactors reflects shutdown of six small demonstration reactors in period 1965-70

Figure 8–13.   Trends in Nuclear Power Plant Licensing

Data for 1955–72 from "Annual Report to Congress," U.S. Atomic Energy Commission, 1972, p. 3. Data for 1973 from "The Nuclear Industry 1973," U.S. Atomic Energy Commission, 1974, pp. 4–7.

producing electricity. Another 56 units were in various stages of construction and 56 more were under review for construction permits.

The 40 units actually operating at the end of 1973 had a total capacity of 24,000 megawatts which represented 5.5% of the total electrical power generating capacity of the U.S. According to current AEC estimates, 33% of all electricity generated in this country in 1985 will come from nuclear power plants. By the end of the century, the nuclear contribution is projected to increase to 60%. Table 8—5 contains a forecast of future domestic nuclear energy capacity. Ranges are given because recent experience shows that construction of nuclear generating facilities is often behind schedule igure 8—14 shows how nuclear electrical generating capacity is projected to increase relative to other sources of electrical power.

The contribution of the nuclear energy industry to the total U.S. energy demand is still small in spite of the rapid development just noted. Nuclear electricity produced in 1973 was equivalent to 852 trillion Btu, which is only 1.1% of the total energy used. By 1985 the contribution to total energy usage will still be less than 5%.

Nuclear power generating facilities, both planned and operating, are not uniformly distributed across the United States (see Figure 8—15). As expected, they are concentrated in the highly populated and industrialized areas of the country.

Table 8—5. FORECAST OF DOMESTIC NUCLEAR ENERGY PRODUCTION

Year	Energy Additions During Year (gigawatts)	Cumulative Total (gigawatts)
1975	7.9—17.1	52.1—56.9
1976	3.7—6.2	55.8—63.1
1977	8.8—12.0	64.6—75.1
1978	10.8—21.5	75.4—96.6
1979	20.2—22.3	97.7—116.8
1980	27.2—29.3	127.0—144.0
1981	24.0—28.0	151.0—172.0
1982	22—33	173—205
1983	26—37	199—242
1984	29—44	228—286
1985	28—46	256—332
1990	32—60	412—602
1995	41—82	602—972
2000	43—122	825—1500

From U.S. Atomic Energy Commission as quoted in *Coal Age*, April 1974, p. 80.
Reprinted with permission of McGraw-Hill, Inc.

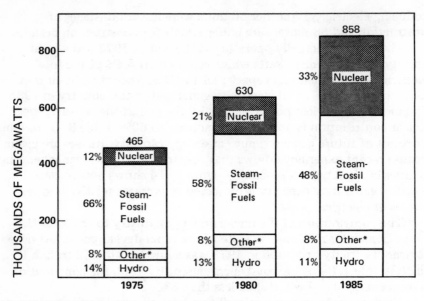

* Includes gas-turbine and internal combustion electrical power generation.

Figure 8–14.   Domestic Electricity Production According to Source. Note that nuclear energy is the only source of electricity projected to increase significantly during the 10-year period included.

Data from "The Nuclear Industry 1973," U.S. Atomic Energy Commission Publication WASH 1174–73, Government Printing Office, 1974, p. 1.

Figure 8–15.   Nuclear Power Reactors in the United States, December 31, 1972

Redrawn from "The Nuclear Industry 1973," U.S. Atomic Energy Commission Publication WASH 1174–73, Government Printing Office, 1974, p. 3.

In many countries nuclear electricity represents a larger percentage of total electricity generated than it does in the United States. Figure 8—16 shows the comparison between the U.S. and 12 other countries, using the common base of 1972 data. Although U.S. nuclear capacity has increased since that time, so has that of other countries.

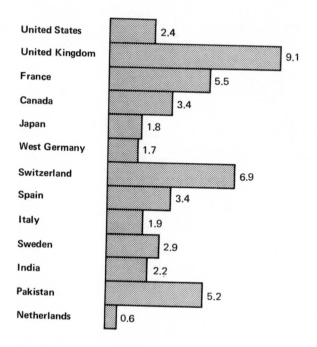

Figure 8—16. Percentage of Total Electricity Produced by Nuclear Power Plants, April 1972

Redrawn with permission from *Fission Reactors: Atoms for Peace Revisited* by Karl P. Cohen (San Jose, Calif.: General Electric Company, 1972), p. 2.

The nuclear power generating facilities that are operating, under construction, or planned for completion in the early 1980s are given in Table 8—6 for some foreign nations. It is quite apparent that nuclear energy development is a worldwide activity. Notice that most of these reactors are of the light water type (BWRs and PWRs) for which the technology is most advanced; only a few are breeders which are still being technologically developed.

Fuels for Fission Reactors

As we have seen, all fission reactors ultimately depend on uranium-235 as a fuel. Even though plutonium-239 and uranium-233

Table 8–6.   FOREIGN DEVELOPMENT OF NUCLEAR POWER FACILITIES

Nation	Number of Reactors by Type							
	BWR	PWR	HTGR	HWR*	PHWR†	FBR**	LMFB	Other
Argentina								
Austria	1				1			
Belgium		4						
Brazil		1						
Bulgaria		4						
Canada	1			1	6			
Czechoslovakia		4						
Finland	1	2						
France						1	1	8
West Germany	10	8	2		1	1		1
East Germany		3						
Hungary		2						
India	3							
Italy	2	1		1		1		2
Japan	13	9				2		2
Korea		1						
Mexico	1							
Netherlands	1	1						
Norway				1				
Pakistan				1				
Spain	2	7						1
Sweden	7	3			1			
Switzerland	3	3						1
Taiwan	4							
United Kingdom			1			2		38
USSR	2	8				4		8
Total	51	61	3	4	9	11	1	61

*Heavy water reactor.
†Pressurized heavy water reactor.
**Fast-breeder reactor.
Data from *Nuclear Engineering International*, April 1973.

are fissionable, they must be produced in breeding reactions that originally began with uranium-235. These produced fuels can be used in the same way as uranium-235 to breed new fuels.

Unlike the fossil fuels, uranium must be put through a complicated process before it can be used as a nuclear fuel. This process, known as a *fuel cycle*, consists of eight steps:

1. Uranium-bearing ore is mined and milled.
2. Uranium oxide is extracted from the ore, refined, and chemically converted into gaseous uranium hexafluoride (UF_6).
3. The uranium hexafluoride is enriched in uranium-235.
4. The enriched uranium hexafluoride is converted into appropriate fuel material, usually uranium oxide or carbide.
5. Fuel elements are fabricated from fuel material.
6. The fuel elements are used in a reactor.
7. Spent (poisoned) fuel elements are reprocessed.
8. The radioactive wastes must be transported and disposed of safely.

Figure 8—17 is a diagram of a nuclear fuel cycle that includes a plutonium recycle step. In this step, plutonium produced during the reactor fission reactions (reaction 8—13) is extracted, converted into a suitable fuel material, and included as a part of the reactor fuel. This is an important process for breeder reactor operations where the amount of new fissionable fuel produced is sufficient to provide fuel for other reactors as well.

Uranium Enrichment

Uranium enrichment (step 3 in the fuel cycle) involves increasing the uranium-235 content of the fuel from its natural level of 0.72% to between 2.5% and 5%. All uranium enrichment in the U.S. is performed in one of three government-owned facilities located at Oak Ridge, Tennessee; Paducah, Kentucky; and Portsmouth, Ohio. These three plants represent a total investment of $2.3 billion. The complexity, construction costs, and operational costs of the enrichment process are undoubtedly factors that keep private industry out of the uranium enrichment business despite a 1971 AEC policy change permitting their participation.

All three United States plants employ the gaseous diffusion process in which gaseous UF_6 is allowed to diffuse through porous barriers. The less massive $^{235}UF_6$ diffuses slightly faster than the heavier $^{238}UF_6$. Therefore, the gas that diffuses through a barrier is slightly richer in the uranium-235 compound than the starting gas. The process is repeated many times on the enriched gas from each previous stage until the desired concentration of uranium-235 is obtained. The gas which is left behind and is lower in uranium-235 is known as the *tails* and is stored to be used as fertile material in breeder reactors. A typical diffusion stage used in the process is shown schematically in Figure 8—18.

The capacity of an enrichment plant is measured in separative work units, SWU. This unit is a measurement of the capability of the plant to perform a specific amount of enrichment under specific conditions. For example, the same number of separative work units can

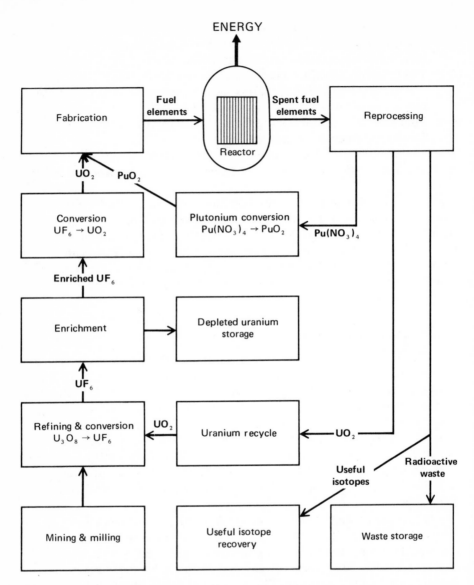

Figure 8—17.   Nuclear Fuel Cycle

produce a large amount of slightly enriched material or a smaller amount of more highly enriched material under the same conditions. The conditions involved are the final concentrations desired, the concentration of feed material, and the concentration of the tails.

The gas centrifuge technique, a second enrichment process, is now under development. The AEC is constructing a pilot plant facility at Oak Ridge, Tennessee, which is expected to be operating by the middle of 1976. One goal of this plant is to compare the economics of gaseous diffusion to gas centrifuge processes.

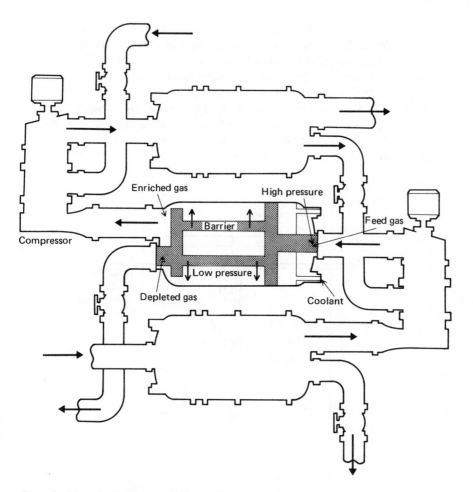

Figure 8—18. Typical Gaseous Diffusion Stage

Redrawn with permission from "Supplying Enriched Uranium" by Vincent V. Abajian and Alan M. Fishman in *Physics Today* (August 1973): 25, left. Reprinted with permission from the American Institute of Physics, Inc.

Gas centrifuge enrichment was attempted during the early stages of atomic bomb development, but technical problems forced its abandonment. Many of these problems were overcome by a new approach to centrifuge construction developed in the 1950s by an Austrian, Gernot Zippe. His design was further improved, and a final report of the work published in 1960 renewed interest in the gas centrifuge technique.

The centrifuge is based on the principle that heavier gas molecules will be forced toward the outer wall of a rapidly spinning container. This leaves a higher concentration of less massive (uranium-235-containing) molecules nearer the center of the container where they can be drawn off. In practice, a mixing action takes place, and enriched gas is drawn off at some distance from the center. The gas centrifuge, shown in Figure 8—19, has the advantage of a larger separation factor than the diffusion method, but it has a lower gas flow rate. For this reason, a large number of centrifuge stages must be used in parallel to produce reasonable quantities of enriched product.

Most of the data concerning the two processes are classified but indications are that the centrifuge method is more economical by as much as 25%. The primary cost advantage is in the area of power needed to run the facility; this necessary power is only 15—17% that of

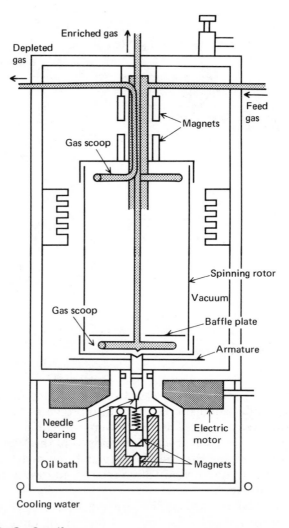

Figure 8—19.   The Gas Centrifuge

Redrawn with permission from "Supplying Enriched Uranium" by Vincent V. Abajian and Alan M. Fishman in *Physics Today* (August 1973): 25, right. Reprinted with permission from the American Institute of Physics, Inc.

a diffusion plant. Centrifuge plants also present less of a hazard in accidents because of the smaller amount of UF_6 feed gas present in the system at any time.

Demand for Fissionable Fuels

According to the data of Figure 8—20, the annual production of uranium oxide peaked in 1960, decreased to a low point in 1966, and has been slowly increasing since then. The resulting supply of enriched

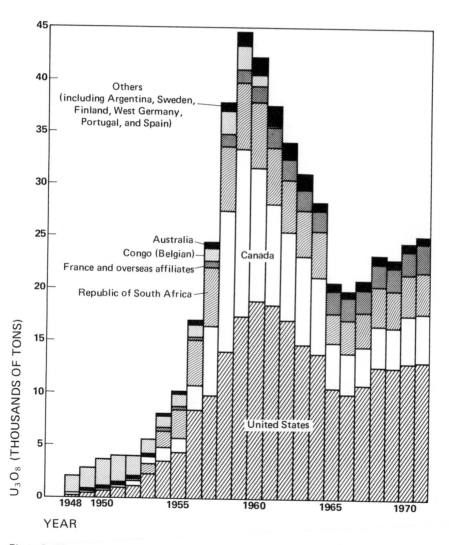

Figure 8—20.   Uranium Oxide Production

From "Nuclear Energy Resources, A Geologic Perspective," U.S. Dept. of the Interior, USGS: INF-73-11, p. 6.

uranium was more than adequate to meet both domestic and world demands through the early 1970s. However, the situation is expected to change about 1977 when increasing domestic and world demands are projected to exceed the available U.S. diffusion-enrichment capacity. Anticipated improvements in the AEC-owned diffusion plants will put this off until about 1979. Domestic demands alone will not exceed the improved capacity until about 1983. However, it is apparent that unless new enrichment capacity is added by 1980, nuclear fuel supplies will be adversely affected.

The development and use of breeder reactors will dramatically influence future demands for uranium as shown by Figure 8—21. This

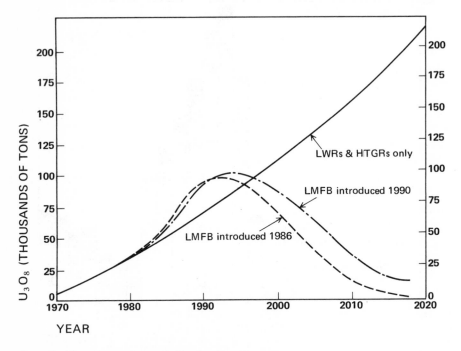

Figure 8—21.   Annual Domestic Uranium Requirements

From "Outlook for Uranium Production to Meet Future Nuclear-Fuel Needs in the United States of America," by R. L. Faulkner in *Peaceful Uses of Atomic Energy*, vol. 8, Proceedings of the International Atomic Energy Agency, 1972, p. 34.

figure shows projected domestic uranium oxide requirements on the basis of three situations: (1) only LWR and HTGR reactors are used, (2) LMFB reactors are introduced by 1986, and (3) LMFB reactors are introduced by 1990. The impact of breeders on demand is obvious. Breeder use significantly increases uranium oxide demand for a time (1986—94) because of the requirement for fertile uranium-238 compounds. The pronounced decrease in uranium demands after the middle 1990s when breeders are used is a dramatic representation of their efficiencies. As much as 70% of the fission energy of natural uranium ($^{238}_{92}$U) is recovered by breeders while light water reactors are capable of extracting only about 2%.

During the middle and late 1970s, the AEC, private industry, and foreign nations will also require substantial amounts of plutonium. The plutonium will be needed for research and development work concerned with reactor assemblies, experimental reactors, and demonstration reactors. Tables 8—7 and 8—8 contain data that show, respectively, the projected needs and available supplies of plutonium. It appears likely that plutonium needs can be supplied well into the 1980s.

Fissionable Fuel Reserves

It should be remembered that uranium ore is mined, milled, and concentrated into U_3O_8 by private industry, which is based on a profit motive. In addition, much of the uranium or other fissionable fuels produced in the future will be purchased by private companies whose activities are also influenced by expenses. For these and other similar reasons, the reserves of fissionable fuels are estimated on a cost basis. The reserves are most commonly listed in terms of cost of production per pound of material. In the case of uranium, cost categories of $8, $10, $15, and $30 per pound of U_3O_8 are used.

Reserves are also classified as being *reasonably assured resources* or *estimated additional resources.* Reasonably assured resources are those contained in known ore deposits that can be recovered using existing technology. Estimated additional resources are contained in ore deposits that are projected to be found in unexplored regions of known deposits or in deposits not yet discovered but located in developed areas.

Table 8—7. ANNUAL PLUTONIUM REQUIREMENTS BY PROGRAM
(kilograms total plutonium)

Program or Project	Year 1975	1976	1977	1978	1979	1980
Fast Flux Test Project (core requirements)	0	0	920	800	800	800
LMFB Reactor (Demo)	0	1310	1310	1740	1740	1740
LMFB Reactor (R & D, fuel technology)	60	60	50	60	60	60
Other	500	0	0	0	0	0
Total	**560**	**1370**	**2280**	**2600**	**2600**	**2600**

Adapted from "The Nuclear Industry 1971," U.S. Atomic Energy Commission Publication WASH 1174—71, Government Printing Office, 1972, p. 49.

Table 8—8.   ANTICIPATED RECOVERY OF PLUTONIUM FROM
DOMESTIC POWER FUELS
(kilograms plutonium per year)

Year	United States	Other Free-World Countries
1971	400	3,500
1972	500	3,800
1973	900	4,000
1974	2,100	4,500
1975	4,000	5,500
1976	6,400	6,700
1977	8,900	8,400
1978	10,800	10,600
1979	12,400	12,900
1980	15,600	16,200
1981	19,300	19,800
1982	22,500	23,800
1983	27,400	27,600
1984	32,300	32,200
1985	37,100	37,400

From "The Nuclear Industry 1971," U.S. Atomic Energy Commission
Publication WASH 1174—71, Government Printing Office, 1972, p. 48.

Uranium reserves are also found as by-products formed during the
processing of other natural resources such as copper ore and phosphate
rock. Israel, for example, contains no proven uranium deposits but does
have phosphate rock deposits totalling about 220 million tons. This
rock, upon processing, would yield about 25,000 tons of uranium. In
the United States, as much as one third of all uranium reserves may
consist of by-products.

Table 8—9 contains estimates of domestic uranium resources.
Total reserves with production costs under $10 per pound of U_3O_8 are
well in excess of a million tons, an amount that should be adequate to
supply demand through the 1980s with or without the introduction of
breeder reactors (see Figure 8—21). Most of the lower priced, reasonably assured reserves are located in the western United States, as shown
by Figure 8—22. It should be noted that the 1974 price for U_3O_8
production is slightly over $6.00 per pound.

Problems Associated with Nuclear
Energy Use

The primary problems associated with the generation and use of
nuclear electricity fall into the categories of environmental damage and

Table 8–9.   UNITED STATES URANIUM RESOURCES,
JANUARY 1, 1974 (tons of U_3O_8, cumulative)

Cost of Production ($ per pound)	Reasonably Assured (proven reserves)	Estimated Additional (potential reserves)	Total
$8 (or less)	275,000	450,000	725,000
$10 (or less)	430,000*	700,000	1,130,000
$15 (or less)	525,000*	1,000,000	1,525,000
$30 (or less)	700,000*	1,700,000	2,400,000

*Includes 90,000 tons of potentially recoverable U_3O_8 as a by-product of phosphate and copper mining at a cost of $10 per pound or less.
Data from Atomic Energy Commission as published in *Coal Age*, April 1974, p. 80. Reprinted with permission of McGraw-Hill, Inc.

safety. Few aspects of the nuclear power program have drawn more criticism and concern than the hazards associated with reactor operation and the handling of highly radioactive materials. In numerous instances construction timetables and plant operating levels of generating facilities have been influenced significantly as a result of public hearings and decisions growing out of this concern. The general categories of problems creating the concern are:

1.   Environmental pollution by radioactive materials during normal operations
2.   Thermal pollution of natural waters during normal operation
3.   Disposal of highly radioactive wastes produced during normal operations
4.   The potential for accidents resulting in the release of massive amounts of highly radioactive materials

With the exception of thermal pollution, all of these problems ultimately focus on the potential danger of exposing living organisms to radiation. Biological effects resulting from such exposure vary and depend on a number of factors, including the total dose of radiation received, the time period involved, the type of radiation, and whether the radiation is received internally or externally. Table 8–10 shows the

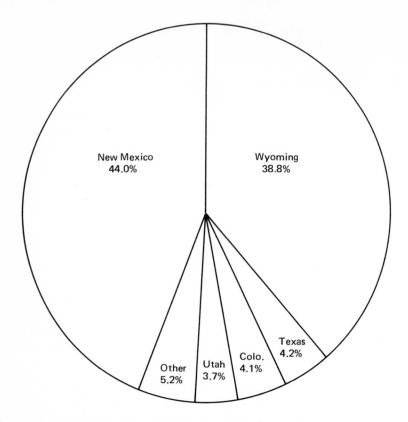

Figure 8—22.　Distribution of Domestic Uranium Reserves Costing $10 or Less per Pound U_3O_8

Data from "The Nuclear Industry 1971," U.S. Atomic Energy Commission Publication WASH 1174—71, Government Printing Office, 1972, Table I-3, p. 17.

effects of whole-body radiation doses received over a short time. The dosage is given in terms of the rem, a commonly used unit that is related to the damaging effects of radiation on tissue. A comparison of these data with the data of Table 8—11 provides some perspective concerning the hazards associated with normal nuclear reactor operations. Of course, increases in the number of reactors or nuclear accidents could increase the hazards, especially in localized areas.

Environmental Pollution

The release of radioactive substances into both the atmosphere and natural waters seems to be an unavoidable consequence of normal activities related to the use of nuclear energy. The mining, milling, and fabricating of nuclear fuels generate only insignificant amounts of radioactive pollutants on a nationwide basis. However, problems have arisen in areas located near facilities engaged in such operations. It is

Table 8—10.   EFFECTS OF RADIATION ON HUMANS FROM WHOLE-BODY,
SHORT-TERM EXPOSURES

Dose (rems)	Effects
0—25	No detectable clinical effects.
25—100	Slight short-term reduction in number of some blood cells, disabling sickness not common.
100—200	Nausea and fatigue, vomiting if dose is greater than 125 rems, longer-term reduction in number of some blood cells.
200—300	Nausea and vomiting first day of exposure, up to a two-week latent period followed by appetite loss, general malaise, sore throat, pallor, diarrhea, and moderate emaciation. Recovery in about three months unless complicated by infection or injury.
300—600	Nausea, vomiting, and diarrhea in first few hours. Up to a one-week latent period followed by loss of appetite, fever and general malaise in the second week, followed by hemorrhage, inflammation of mouth and throat, diarrhea, and emaciation. Some deaths in two to six weeks. Eventual death for 50% if exposure is above 450 rems; others recover in about six months.
600 or more	Nausea, vomiting, and diarrhea in first few hours. Rapid emaciation and death as early as second week. Eventual death of nearly 100%.

From *Medical Aspects of Radiation Accidents*, E. L. Saenger, ed. (Washington, D.C.: U.S. Atomic Energy Commission, 1963), p. 9.

the opinion of most experts in the field that the enforcement of regulations related to the location and storage of mine and mill tailings and waste solution ponds can adequately solve such problems.

During operation, nuclear power plants produce a variety of radioactive materials that are released as liquids or gases, depending on the type and design of the reactor involved. The principal radioactive materials formed are fission products — the ashes of the nuclear fuel. Other radioactive materials are generated when neutrons bombard small amounts of impurities found in the circulating coolant. Some of these impurities are found in the coolant as a result of the limits imposed by purification techniques; others are the products of the corrosion and erosion of reactor structural materials.

The gaseous radioactive products from boiling water and pressurized water reactors are mainly short-lived isotopes of the rare gases. In the pressurized water reactor, few are released into the environment because they are retained in the primary circulating coolant system

Table 8–11.   AVERAGE ANNUAL RADIATION DOSE EQUIVALENTS RECEIVED BY
THE GENERAL POPULATION

Dose (rems)	Source
0.100–0.140	Natural background radiation
0.020–0.050	Diagnostic X rays
0.003–0.005	Therapeutic X rays
0.0002	Medical radioisotopes
0.0015	Fallout (from atmospheric testing, 1954–62)
0.00085	Radioactive pollutants from nuclear power plants (1970)
0.002	Smoking of one pack of cigarettes per day
0.170	Maximum annual limit proposed (by the International Commission on Radiological Protection, 1970) from all sources exclusive of medical and background

Reprinted with permission from *Elements of General and Biological Chemistry* by John F. Holum (New York: John Wiley and Sons, Inc., 1972), p. 535.

where they rapidly decay. However, in boiling water power plants, the radioactive gases pass (with the steam) from the reactor directly to the turbine. The steam is then changed into a liquid in the condenser, but the rare gases remain in the gaseous state and are removed by an air ejector. The ejected radioactive gases are circulated and allowed to decay before being released into the atmosphere. Circulation times of 30 minutes are often used, but the hazards can be reduced much more by holding the gas several days. Under these conditions, the radioactive emissions would be limited to the long-lived krypton-85 which is present in concentrations considered to be insignificant. In addition to holdup techniques, effluent gases are also subjected to filtration designed to remove radioactive particulates produced during the decay of radioactive gases.

In contrast to gaseous radioactive substances, those produced in liquid form are long-lived. Holdup and decay techniques are applied to these materials, but the resulting decrease in radioactivity is slight. For a typical situation, holdup times of forty days would decrease radioactivity levels only by a factor of five. Filtration is the sole method used to remove insoluble and particulate matter from liquid wastes. Additional treatment steps are evaporation and ion-exchange demineralization. Evaporation is used to concentrate radioactive substances into a smaller volume, which is then treated by ion-exchange methods or removed to off-site disposal areas. Ion-exchange treatment removes and retains radioactive ions in much the same way a water softener removes and traps hard-water ions. After appropriate combinations of these treatment processes have been used, liquid wastes are discharged into natural waters.

The procedures described above result in the routine discharge of slightly radioactive effluents into the environment. This practice has received criticism based on several possibilities:

1. The allowable effluent radiation levels are based on radiation effects that could be much more severe than now believed.
2. The radiation exposures resulting from the radioactive effluents could be much greater than those calculated and thought to be correct.
3. Allowable effluent radiation levels are based on radiation effects which are believed to occur infrequently. The assumptions on which this belief is based could be in error.

Further research will hopefully resolve some of these issues. Until that time, it is comforting to realize that the actual radiation levels of effluents from nuclear power plants are usually less than 10% of the allowable maximum.

It has been stated before that thermal pollution of water is at the present an unavoidable result of energy generation involving a thermal cycle. This is a particularly significant problem with nuclear power generation because present light water reactors discharge about 50% more waste heat than do modern fossil-fueled plants. The improvement of light water reactor efficiencies and the use of the more efficient high-temperature gas-cooled reactors will alleviate the problem to some degree. However, a complete solution will result only from further research into the use of air cooling towers or alternate approaches.

Radioactive Waste Disposal

Some nuclear liquid and solid wastes are too highly radioactive to be safely released into the environment. These materials are classified as high- or low-level wastes. High-level wastes are those resulting from the fuel purification steps of a fuel reprocessing plant, or any material with an equivalent radioactivity. Wastes not classified as high level but considered to be unsuitable for release into the environment are classified as low level.

Low-level wastes include such diverse materials and items as drain water, decontamination wash water, filter elements, spent ion-exchange resins, paper, rags, clothing, and tools. The dry wastes are collected, compressed and, if possible, sealed in 55-gallon drums/ The wet solids and liquid concentrates are mixed with cement or other immobilizing materials and also sealed in drums. Most of this material is then shipped to appropriate sites for near-surface burial. This form of disposal is carried out by commercial companies and the Atomic Energy Commission. The annual amount of low-level waste buried by the AEC has been fairly constant since 1968 at between 1.3 and 2.0 million cubic feet. The amounts buried at commercial sites have been steadily increasing, as shown by the data of Table 8—12, which gives both commercial and AEC burial amounts.

Table 8—12.   ANNUAL BURIAL OF LOW-LEVEL WASTES
(thousands of cubic feet of solid waste)

Year	Amount Buried Commercial Sites	AEC Facilities	Total
1968	666.6	1,747	2,413.6
1969	750.9	1,961	2,711.9
1970	995.1	1,650	2,645.1
1971	1,206.8	1,403	2,609.8
1972	1,334.5	1,407	2,741.5
1973	1,500*	1,300*	2,800
1974	2,000*		
1975	4,000*		
1980	6,000*		

*Estimates by AEC.

Data from "The Nuclear Industry 1973," U.S. Atomic Energy
Commission Publication WASH 1174—73, Government Printing Office,
1974, pp. 59, 60.

High-level wastes are handled differently. They are now stored as
liquids in large underground tanks. However, according to an AEC
policy that became effective in 1971, all high-level wastes must be
converted into solid form within 5 years of their production and
shipped to a federal repository within 10 years of production. More
than 80 million gallons of high-level wastes are currently stored in
nearly 200 tanks ranging in volume from 300 thousand to 1.3 million
gallons. These tanks are located at the AEC's Hanford, Idaho Falls, and
Savannah River sites. Some of the older tanks (more than 20 years old)
have leaked and released radioactive materials into the ground. Despite
the contention by the AEC that no hazards were created by the leaks, a
significant number of individuals and organizations have expressed
concern. It has been pointed out that little data exists to indicate just
how rapidly the leaked material might reach underground water
systems or the effect such material might have on other ecological
systems. It has also been noted that (certainly hazardous) massive leaks
could occur if the storage tanks were subjected to earthquakes or other
similar natural stresses.

Technical problems, plus projections indicating an increasing
future production of high-level wastes, make it imperative that disposal
methods other than tank storage be developed. All of the proposals now
being studied involve the conversion of liquid wastes into solids prior to
storage. The solidification immobilizes the wastes and reduces their
volume by 90%. Table 8—13 contains estimates of the future quantities
of wastes needing disposal and shows the effect of solidification on volume.

Table 8–13. ESTIMATED HIGH-LEVEL WASTE PRODUCTION FROM THE NUCLEAR POWER INDUSTRY

	Year			
	1970	1980	1990	2000
Installed generating capacity (thousands of megawatts)	6	150	450	940
Reprocessed fuel per year (tons)	55	3,000	9,000	19,000
Annual production of high-level liquid wastes (thousands of gallons)	17	970	3,300	5,800
Accumulated high-level wastes if not solidified (thousands of gallons)	400	4,400	29,000	77,000
Annual production of solidified wastes (thousands of cubic feet)	—	9.7	33	58
Accumulated volume of solidified wastes (thousands of cubic feet)	—	44	290	770

Reprinted with permission from *Engineering for Resolution of the Energy-Environment Dilemma,* Publication ISBN 0-309-01943-5, Committee on Power Plant Siting, National Academy of Engineering, Washington, D.C., 1972, p. 205.

In 1970, the AEC announced the tentative selection of a salt mine near Lyons, Kansas, as the initial site for long-term storage of solidified commercial high-level wastes. A salt mine was chosen because of salt characteristics and geological factors associated with salt mines that appeared to be very suitable for the projected long-term storage. An artist's drawing of the project is given in Figure 8–23. The proposal created a great deal of controversy and many questions were raised. In May 1972 the AEC announced that the project would be held up pending further study because many of the questions could not be resolved in a reasonable time.

An alternative proposed by the AEC involves surface storage of solids in a Retrievable Surface Storage Facility (RSSF). The reference design for the RSSF utilizes a water-basin concept of storage in which waste-containing cannisters are stored in water-filled, stainless steel-lined basins. Heat from radioactive decay is to be rejected to the atmosphere by a system of heat exchangers and cooling towers. The RSSF facility is expected to be ready for use by 1983.

Other RSSF storage concepts are also being evaluated. For example, an air-cooled vault approach is one alternative. Heat is removed by pumping air through the storage vault and out an exhaust port. The salt mine idea is also still under consideration because such storage facilities require less maintenance and monitoring than RSSF storage. A Bedded Salt Pilot Plant (BSPP) project is under way with the goal of resolving the earlier objections to the salt mine project.

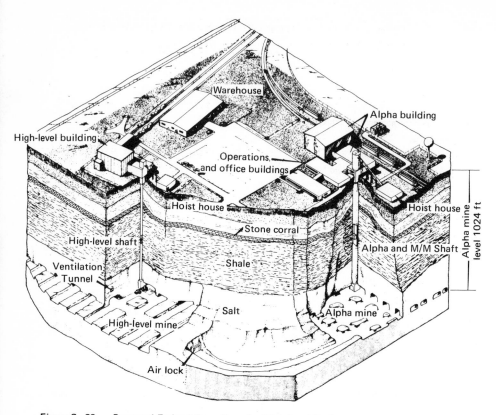

Figure 8–23.   Proposed Federal Repository for High-Level Radioactive Wastes

From "Current Developments in Long-Term Radioactive Waste Management," by R. L. Culler, J. O. Blomeke, and W. G. Belter in *Peaceful Uses of Atomic Energy*, vol. 11, Proceedings of the International Atomic Energy Agency, 1972, p. 434.

Accidental Release of Radioactive Materials

The potential for accidents is present in many of the procedures associated with nuclear energy use. Particular concern has been expressed about the transporting of radioactive materials and the operation of reactors, since large amounts of very radioactive materials are involved. Accidents during transport could possibly result in the release of dangerous substances in highly populated areas through which highways or railroads pass. Even though nuclear power plants are often located a significant distance from population centers (a practice that may become impractical as the number of plants is increased), an extreme accident could scatter highly radioactive materials far enough from the plant to reach the populated areas with disastrous results.

The Atomic Energy Commission and the Department of Transportation are jointly responsible for protecting the health and safety of the public against potential hazards created by the transporting of radioactive materials over public carrier systems. It is obviously impossible to completely eliminate the chances for a carrying vehicle to be involved in an accident. For this reason, most safety regulations are aimed at providing containers with the ability to retain their integrity under the most severe conditions likely to be encountered. The transported materials may be nuclear fuel elements, enriched raw materials destined to be fabricated into fuel elements, spent fuels to be reprocessed, high-level wastes, or other miscellaneous materials. The heavy radiation and heat shielding required together with the wide diversity of shapes, sizes, and weights involved, create some distinct challenges for container designers. However, the resulting systems are generally acknowledged to be very satisfactory and development work continues as new needs are discovered.

An operating nuclear reactor contains huge quantities of radioactive substances that must be kept under control during normal operations and also when accidents occur. The precautions taken to control radioactive substances during accidents have become subjects of controversy both within and without the AEC.

Concern is sometimes expressed about the possibility of a fission reactor running wild and becoming a nuclear bomb. Even most critics of the current safety precautions acknowledge that this cannot happen. As the rate of nuclear reactions within a reactor core increases, the temperature also goes up. The increase in temperature has the effect of slowing and stabilizing the reaction.

Another type of accident is the breakdown of fuel cladding integrity. The cladding is essentially a metal tube in which fuel is sealed before being placed in the reactor core. A typical core contains many of these tubes or fuel elements. If the cladding were to fail, fuel and fission products would be released into the primary coolant system which thus behaves as a second containment barrier (the cladding is the first barrier). In the event this second barrier breaks down, a third containment barrier, the structure housing the reactor, is available. Reactor containment structures are designed to retain all radioactive materials released under credible (likely to be encountered) circumstances. Once again, most critics agree that the containment system would probably function under the credible circumstances, but there is some question about the validity of what are assumed to be credible circumstances.

Another potential accident also raises the issue of credible circumstances. This accident, considered to be the most extreme that could occur in light water reactors, would involve the sudden loss of primary reactor coolant. Under the worst credible circumstances, a rupture of all pipes carrying coolant into and out of the reactor is assumed (critics point out that it is not considered credible to imagine a rupture of the reactor vessel itself). Under these conditions of complete coolant loss, heat from the decay of core materials could melt the core, destroy the fuel cladding, and cause a nonnuclear explosion. There is a great deal of disagreement about the ability of containment systems to function properly under such circumstances.

As a precaution against coolant-loss accidents, all commercial light water reactors contain emergency core-cooling systems which would automatically flood the core with coolant in the event of an accident. The effectiveness of such systems is questioned for a number of reasons. For example, some preliminary experiments indicate that emergency cooling systems now installed might not work. In tests conducted by the AEC in 1970 and 1971, small-scale reactor core mock-ups were electrically heated to temperatures approximating those present during a loss-of-coolant accident. Attempts were then made to cool the cores by techniques which simulated the behavior of installed emergency systems. The fuel elements swelled at the high temperatures and blocked the passage of 80—85% of the emergency coolant through the hot core.

The AEC's LOFT (loss-of-fluid test) project could provide solid data on which to evaluate the design and effectiveness of future as well as current emergency core cooling systems. This project is now under way after suffering from many construction delays. Originally, it was planned to allow the reactor of the LOFT facility to self-destruct after total loss of coolant as a final experiment; costs and other considerations have forced cancellation of this experiment. However, the sophisticated and highly instrumented facility is capable of simulating a wide variety of coolant-loss accidents, including all that are considered even remotely possible in commercial power reactors.

It is to be hoped that supporters of current reactor safety procedures are correct when they say that the extreme over-engineering and attention-to-detail that go into the construction of nuclear power plants make the likelihood of accidents very small. It is also not unreasonable to assume that in the event weaknesses are found in safety systems, suitable steps will be taken, such as the lowering of allowable upper limits of power generation from a power plant.

Fusion Reactors

Nuclear fusion is the source of energy which some believe will be the ultimate solution to the energy supply problem. With the exception of solar energy, this source alone has the advantage of a widely distributed, low-cost, and essentially unlimited fuel supply. However, the feasibility of fusion reactors is still to be proven. The estimated time required to prove feasibility varies with a number of factors, including the person doing the estimating. One very important factor in common seems to be the level of financial support given to the researchers. Under favorable financial conditions, feasibility could be proven by the early 1980s and commercial fusion energy could be available by the turn of the century.

Fusion Fuels and Reactions

The light elements deuterium (hydrogen-2), tritium (hydrogen-3), helium-3, and lithium-6 appear to be the only practical fusion fuels available for use. The characteristics of the potentially useful reactions of these substances are summarized in Table 8—14.

The very high ignition temperatures make it unlikely that the last two reactions in Table 8—14 will be used until fusion technology is well advanced — possibly after the year 2000. Similarly, the low ignition temperature of the first reaction makes it a prime candidate for use in the first nuclear reactors in spite of one drawback. Only one of the reactants, deuterium, occurs naturally in the quantities needed for energy production. It is found in seawater (as heavy water) in the proportion of one part in 6000. Tritium, the other reactant, will have to be produced synthetically, but fortunately the fusion reaction provides a way to do this. Neutrons from the fusion reaction can be absorbed into a layer of lithium metal to produce tritium:

$$\,^1_0 n + \,^6_3 Li \longrightarrow \,^3_1 T + \,^4_2 He \qquad\qquad \text{8—15}$$

$$\,^1_0 n + \,^7_3 Li \longrightarrow \,^3_1 T + \,^4_2 He + \,^1_0 n \qquad\qquad \text{8—16}$$

The fusion reactor thus becomes analogous to a fission breeder reactor.

Fusion reaction temperatures are high enough to completely separate the electrons from the nuclei of the light elements undergoing fusion. Thus, as indicated by the reactions of Table 8—14, the reactants and products of fusion reactions are charged nuclei, protons, and neutrons. The high-temperature, gaslike fusion mixture of electrons, nuclei, and protons is called a *plasma* and is considered by some to represent a fourth state of matter (in addition to solids, liquids, and

---

Table 8—14.   POTENTIALLY USEFUL FUSION REACTIONS

Reaction		Energy Released (Btu)	Ignition Temperature (°K)
1.	$\,^2_1 D^+ + \,^3_1 T^+ \longrightarrow \,^4_2 He^{2+} + \,^1_0 n$	26.8×10^{-16}	100 million
2.	$\,^2_1 D^+ + \,^2_1 D^+ \longrightarrow \,^3_2 He^{2+} + \,^1_0 n$	5.0×10^{-16}	600 million
3.	$\,^2_1 D^+ + \,^2_1 D^+ \longrightarrow \,^3_1 T^+ + \,^1_1 H^+$	6.1×10^{-16}	600 million
4.	$\,^2_1 D^+ + \,^3_2 He^{2+} \longrightarrow \,^4_2 He^{2+} + \,^1_1 H^+$	27.8×10^{-16}	1 billion
5.	$P^+ + \,^6_3 Li^{3+} \longrightarrow \,^3_2 He^{2+} + \,^4_2 He^{2+}$	6.1×10^{-16}	2 billion

Data from "Controlled Thermonuclear Research," Part 2, U.S. Congress, Subcommittee Hearings, Nov. 10—11, 1971, p. 275.

gases). The design characteristics of fusion reactors will depend a great deal upon the characteristics and behavior of the plasma involved.

Proof of Feasibility

Opinions differ about what successful experiments will constitute a proof of the scientific feasibility of controlled-fusion energy generation. However, some combination of the following achievements is generally thought to be necessary; all will be required to produce a useful commercial reactor:

1. Attain ignition temperatures
2. Contain the resulting plasma
3. Attain an energy balance
4. Add fuel to the reactor
5. Extract useful energy from the reactor

Ignition temperatures must be attained in order to initiate fusion reactions. In continuous reactors it is assumed that an energy balance will be established and the ignition temperature will be maintained by energy produced from the fusion reactions. The ignition temperatures given in Table 8—14 are not the lowest (ideal) for each reaction. Instead, they are temperatures considered to be practical under the less-than-ideal conditions characteristic of real systems. If more energy is lost from a hot reacting plasma than is generated within the plasma, cooling results. Eventually the plasma temperature will become too low for the fusion reactions to continue. Thus, the attainment and maintenance of ignition temperatures in a plasma are related to the balance between energy production and energy losses in a plasma.

Energy can be removed from plasmas by various mechanisms. For example, neutrons produced by fusion reactions can escape and carry their kinetic energy with them. At fusion temperatures, the major mechanism for plasma energy loss is bremsstrahlung radiation, a continuous emission of electromagnetic radiation that is similar to X rays. Bremsstrahlung radiation results when charged particles such as electrons, nuclei, or protons are accelerated. Heavy nuclei in a plasma greatly increase losses by this mechanism so plasmas must be kept free of such contaminants — a topic discussed later in conjunction with plasma containment. An ideal energy balance is achieved when bremsstrahlung losses and energy generation within the plasma are equal. The temperature at which this occurs is the ideal ignition temperature and is illustrated in Figure 8—24 for the D-T and D-D reactions (reactions 1, 2, and 3 of Table 8—14). Note that an ideal energy balance does not allow for the extraction of useful energy from the reactor. In order to produce energy that could be removed from the reactor and used, the plasma would have to generate more energy than was lost by brems-

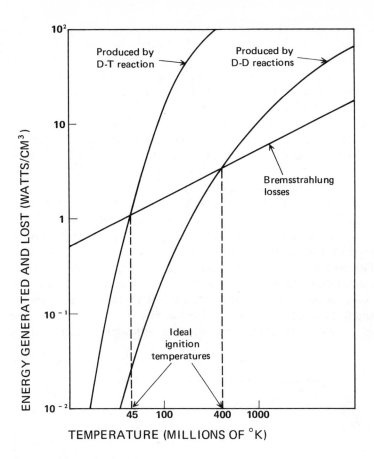

Figure 8—24. **Energy Balance in Fusion Processes**

Adapted from *Controlled Nuclear Fusion* by Samuel Glasstone (Washington, D.C.: U.S. Atomic Energy Commission, n.d.), p. 13.

strahlung radiation (and other less important mechanisms). This can be achieved by operating at higher temperatures, as shown in Figure 8—24, and is one reason for the higher-than-ideal ignition temperatures of Table 8—14.

A further condition related to energy balance was described by J. D. Lawson in 1957. In simplified form, the Lawson criterion sets forth conditions under which the amount of energy that can be extracted from a fusion reaction is just sufficient, if returned to the reactor, to cause the reaction to be self-sustaining. Thus, the minimum Lawson criterion defines the break-even point for fusion reactors. The energy produced by a plasma reaction in a given time depends upon the number of fusions that occur in that time. The number of fusions, in turn, depends upon the chances for nuclear collisions to take place. Increases in nuclear velocities (proportional to temperature) and the number of nuclei contained per unit volume (ion density, n_i) increase the chances for nuclear collisions. The Lawson criterion can best be expressed in terms of the product $n_i t$ where t is the plasma confinement time. On this basis, a dense plasma confined for a short time is equivalent to a less dense plasma confined for a longer time. The Lawson break-even conditions are shown in Figure 8–25, where it is assumed that the energy from the reactor is run through a thermal recovery cycle (heat exchangers, etc.) with an efficiency of 33%. We see that the $n_i t$ product for reactions carried out at 100 million °K has a minimum value of 8×10^{14} for the D-T fuel system. This corresponds to an ion density of 10^{14} to 10^{15} particles per cubic centimeter — about 1/10,000 as dense as the atmosphere — and confinement times of between 0.1 second and 1 second. The progress made toward achieving the Lawson criterion by using various plasma heating and containment systems is also indicated in the figure.

The addition of fuel to a reactor may not prove to be a simple task. Calculations based on present knowledge indicate that attempts to inject fuel droplets into a hot plasma might result in one of two unsatisfactory situations. The fuel could vaporize, ionize, and become trapped on the plasma surface where low temperatures might curtail

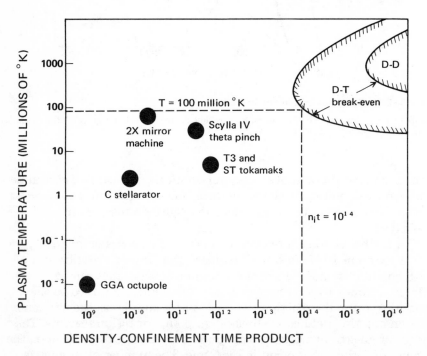

Figure 8–25.   The Lawson Criterion for Thermonuclear Fusion

Reprinted from "Controlled Thermonuclear Research," Part 2, U.S. Congress, Subcommittee Hearings, Nov. 10–11, 1971, p. 277.

fusion reactions, or the droplets could pass completely through the plasma without reacting. More research is needed in this area; if the problem cannot be solved, future reactors, instead of operating continuously, might have to be run in repeated cycles of fuel injection, ignition, and energy extraction.

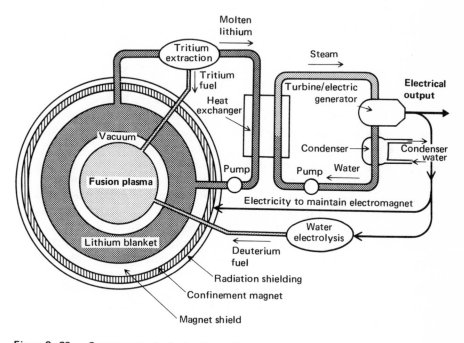

Figure 8—26.   Components of a Fusion Power Plant with Thermal Energy Extraction

The largest part of the energy released in the fusion of deuterium and tritium appears as neutron kinetic energy (see reaction 1 of Table 8—14). A substance circulated around the hot plasma can then be pumped outside the reactor where the heat is removed and used to run a steam turbine electrical generator. Molten lithium metal used as the circulating material has the added advantage of generating the tritium needed in the D-T fuel system (reactions 8—15 and 8—16). The essential components of a fusion power plant of this type are represented schematically in Figure 8—26. As fusion technology is improved, it is contemplated that other higher temperature reactions which release a larger portion of the fusion energy in the form of charged particles will become feasible — reaction 3 of Table 8—14, for example. This type of energy can be extracted directly in the form of electricity by using conversion techniques such as magnetohydrodynamics (see Chapter 9).

Magnetic Plasma Containment Systems

One of the fundamental but difficult problems of fusion power development is that of confining the hot plasma for the reaction times implied by the Lawson criterion. Research toward this end has been going on throughout the world for more than 20 years. The primary problem of plasma confinement is created by the necessity of using a nonmaterial container. This requirement is sometimes erroneously thought to be necessary because the hot plasma would destroy any material it contacted. However, the plasma density is so low that the total energy content would be insufficient to damage container walls. On the contrary, a nonmaterial container is necessary to keep the plasma from the walls to avoid conductive heat losses that would result in plasma temperatures too low to support fusion. In addition, contact with container walls introduces heavy contaminants into the plasma which greatly increase bremsstrahlung radiation losses.

A nonmaterial containment system is made possible by the electrically charged nature of plasmas. Charged particles have difficulty crossing magnetic lines of force, so a magnetic "bottle" can be created by surrounding the plasma with a suitable magnetic field. Magnetic fields of various shapes can be generated by passing an electric current through an appropriately arranged pattern of wires or other conductors.

Magnetic fields may be used for heating plasmas as well as for confinement. This can be accomplished by using a magnetic field to compress a low-temperature plasma. The compression causes a dramatic increase in plasma temperature. Resistive, or ohmic, heating is also used but has not proved to be as effective as magnetic compression. During resistive heating a large current is passed through the conductive plasma which becomes hot in somewhat the same way as the heating element of an electric stove. Various other combinations and variations of magnetic and resistive heating are also used.

Magnetic confinement systems are classified as open or closed and have the fundamental geometries shown in Figure 8—27. In general, charged particles move along in a magnetic field by spiralling around the lines of force. Therefore, in closed systems the particles are confined to move and remain inside the toroidal (donut-shaped) volume containing the magnetic lines of force. In an open system some sort of method must be employed to prevent the particles from following the lines of force through the ends of the container. This is done by creating magnetic mirrors (high-intensity magnetic fields) at the end of the system or magnetic wells (low-intensity areas) in the confinement region within the system. In practice, plasma tends to leak out of both types of confinement systems. A great research effort is devoted to minimizing the leaks to the extent of achieving containment times that will satisfy the Lawson criterion.

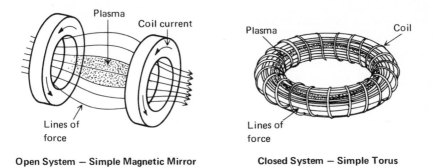

Open System — Simple Magnetic Mirror   Closed System — Simple Torus

Figure 8—27.   Geometries of Open and Closed Magnetic Plasma Containment Systems

Redrawn from "Controlled Thermonuclear Research," Part 2, U.S. Congress, Subcommittee Hearings, Nov. 10—11, 1971, p. 279.

Table 8—15 contains a list of some of the devices and experiments being used to investigate plasma confinement problems. Those representing possible fusion reactor configurations are indicated and their performance can be seen by referring to Figure 8—25. Figure 8—28 contains diagrams representing illustrative open and closed containment systems.

An Alternative to Magnetic Confinement

In one of the newest approaches to controlled nuclear fusion, the magnetic confinement of plasmas is unnecessary. Instead, a reactor mechanically strong enough to withstand the detonation of tiny thermonuclear bombs is envisioned. During operation, tiny solid (frozen) pellets containing deuterium and tritium would be struck by a pulsed, high-power laser beam. The surface of the pellet would ignite, and the fusion reaction would rapidly proceed inward (the pellet implodes). The plasma formed during fusion would expand outward, and energetic neutrons absorbed in a circulating layer of molten lithium would provide a means for extracting useful energy and generating the necessary tritium fuel. The essentials of such a reactor system are illustrated in Figure 8—29.

One of the biggest technological problems associated with this concept is the development of the necessary high-power lasers. The most powerful lasers now available generate only about 1% of the power estimated to be needed for the operation of a pellet reactor. However, good progress has been made in laser research during the last decade and a concerted, well-financed effort could result in the necessary development.

Table 8—15.   MAJOR CONTROLLED FUSION EXPERIMENTS

Closed Configurations

Device Name	Experiment Location
Astron	U.S.
Floating rings	France, Germany, Japan, Sweden, U.K., U.S.
Heliotron	Japan
RF interaction	USSR
Stellerator*	France, Germany, Japan, U.K., U.S., USSR
Tokamak*	Australia, Japan, U.S., USSR
Toroidal screw pinch	Netherlands, USSR
Toroidal theta pinch*	Germany, France, U.K., U.S.
Toroidal Z pinch	France, U.K., U.S., USSR

Open Configurations

Device Name	Experiment Location
Cusps	Poland, USSR
Linear theta pinch	France, Germany, Italy, Japan, U.K., U.S.
Linear Z pinch	Italy, U.K., U.S., USSR
Magnetic mirrors*	France, Japan, U.K., U.S., USSR
RF interaction	USSR

*Possible fusion reactor configurations.
Reprinted from "The Search for Fusion Power" by CDR. G. J. Mischke in *Naval Research Reviews*, vol. 24, no. 4 (April 1971): 15.

Advantages and Disadvantages of Fusion Power

Nuclear fusion as an energy source has both advantages and disadvantages when compared to other sources. Some advantages that have been identified are given below.

1. An abundant and widely distributed fuel supply is available — Deuterium, obtained by the electrolysis of seawater, is available in potentially unlimited quantities and could ultimately become the only fuel needed. Until such time, the other fuel, tritium, can be produced by the fusion reactor itself.
2. The radiation hazard is small — During operation, a fusion reactor will be a strong source of radiation. Proper shielding will solve this problem. No radioactive waste materials are generated by fusion reactors. The only radioactive materials involved are the tritium fuel and reactor structures made radioactive by neutron bombard-

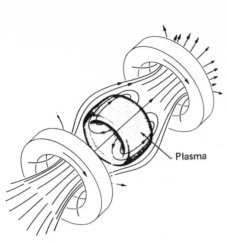

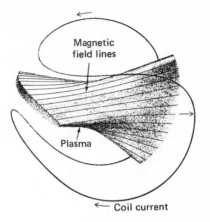

Astron — a magnetic mirror device

OPEN SYSTEMS

"Baseball" coil — current through the
baseball-seam-shaped coil contains
the plasma in a magnetic well

CLOSED SYSTEMS

Tokamak — basic design

Magnet coils
Vacuum chamber
Plasma
Transformer

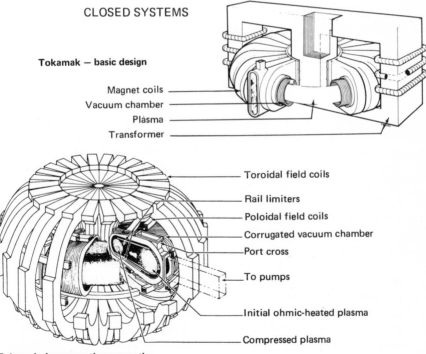

Toroidal field coils

Rail limiters

Poloidal field coils

Corrugated vacuum chamber

Port cross

To pumps

Initial ohmic-heated plasma

Compressed plasma

Tokamak, incorporating magnetic
compression and heating of plasma

Figure 8—28.   Illustrative Plasma Containment Systems

Redrawn from "Controlled Thermonuclear Research," Part 2, U.S. Congress, Subcommittee
Hearings, Nov. 10—11, 1971, pp. 307, 349, 520, and 626.

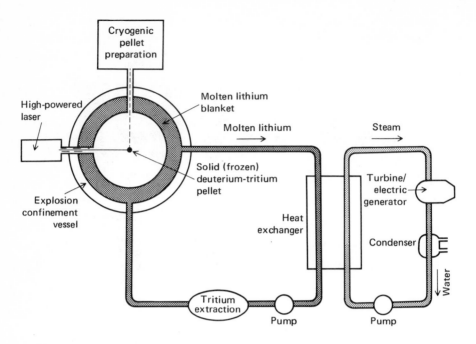

Figure 8—29.   Nuclear Fusion in a Pellet Reactor

ment. Radioactive tritium will be present in high concentrations only in small, confined areas that can be easily protected at low cost. When a fusion plant needs repairs or becomes obsolete and must be replaced, radioactive internal components will have to be handled carefully. They will probably be removed, compressed, and stored until the radioactivity decreases to safe levels. However, studies show that such repairs or replacements should be needed only infrequently, possibly only every 20 years.

3.   Fusion reactors are inherently safe — A fusion reactor cannot undergo runaway reactions. At any time, sufficient fuel is available for only a few seconds of operation. If too much fuel is injected, the plasma is cooled and the reaction stops. If the magnetic confinement system fails, the plasma leaks to the container walls, where it is cooled and the reaction stops. During such an occurrence, the container walls would increase in temperature by no more than 100°C.

4.   Fusion reactors will have high efficiencies — Fusion power plants will probably operate with overall efficiencies as high as 60%, a decided improvement over conventional systems. However, there are a number of serious disadvantages:

1. Feasibility is still to be proven — We must not forget that controlled nuclear fusion may not be possible. If this proves to be the case, nuclear fission breeder reactors will, in the opinion of many, have to provide most future electrical power.

2. Engineering problems will develop — Even after (if) feasibility is proven, the development of commercial reactors will require the solution of a number of problems. The behavior of reactor materials when subjected to huge temperature extremes as well as neutron bombardment could create very challenging engineering and design problems. Other difficult tasks will be the development of appropriate cryogenic magnets, vacuum systems, and heat exchangers.

3. Thermal pollution will be generated — As long as heat is the main form of energy output, thermal pollution from the necessary turbine-generator cycle will be a problem. The development of direct electrical conversion techniques will help diminish this effect.

4. Fusion power tends to come in large packages — The requirement for a heavy lithium radiation shield and blanket to absorb high-energy neutrons makes it uneconomical to build small reactors. The net electrical power output of the smallest economical reactor will probably be in the range of 1000 megawatts.

5. Excessive amounts of some substances will be needed — Estimates are that by the year 2020, at least 1000 fusion reactors will be needed to satisfy electrical energy needs. Unless methods are found to reduce the amount of materials needed, their construction would require more than the available amount of some materials. For example, an estimated 2 million tons of lithium metal would be needed; lithium production at that time is estimated to be about 20 thousand tons per year. Thus, one thousand years production would be required. If the figures are at all accurate, fusion reactor development could not be as rapid as needs would dictate.

Suggested Readings

Abajian, V. V., and A. M. Fishman, "Supplying Enriched Uranium," *Physics Today*, Aug. 1973, pp. 23–29.

Alexander, T., "The Big Blowup Over Nuclear Blowdowns," *Fortune*, May 1973, p. 216.

Blomeke, J. O., and C. W. McClain, "Managing Radioactive Wastes," *Physics Today*, Aug. 1973, pp. 36–42.

Carter, L. J., "Floating Nuclear Plants: Power from the Assembly Line," *Science*, March 15, 1974, pp. 1062–1065.

Culler, F. L., Jr., and W. O. Harms, "Energy from Breeder Reactors," *Physics Today*, May 1972, pp. 28–39.

Department of the Interior, Geological Survey, "Nuclear Energy Resources: A Geologic Perspective," Government Printing Office, Washington, D.C.

Drummond, W. E., "Thermonuclear Fusion: An Energy Source for the Future," *Journal of College Science Teaching*, Oct. 1973, pp. 18–21.

Environmental Protection Agency, Office of Radiation Programs, "Proceedings of Southern Conference on Environmental Radiation Protection from Nuclear Power Plants," Government Printing Office, Washington, D.C., April 21–22, 1971.

Environmental Protection Agency — PB 209283, "Survey of Nuclear Power Supply Prospects," Government Printing Office, Washington, D.C., Feb. 1973.

Forbes, I. A., D. F. Ford, H. W. Kendall, and J. J. Mackenzie, "Cooling Water," *Environment*, Jan./Feb. 1972, p. 40.

Ford, D. F., and H. W. Kendall, "Nuclear Safety," *Environment*, Sept. 1972, p. 2.

Gillette, R., "Nuclear Safety: AEC Report Makes the Best of It," *Science*, Jan. 26, 1973, pp. 360–363.

————, "Nuclear Safety (I): The Roots of Dissent," *Science*, Sept. 1, 1972, pp. 771–776.

————, "Nuclear Safety (II): The Years of Delay," *Science*, Sept. 8, 1972, pp. 867–871.

————, "Nuclear Safety (III): Critics Charge Conflict of Interest," *Science*, Sept. 15, 1972, pp. 970–975.

————, "Nuclear Safety (IV): Barriers to Communication," *Science*, Sept. 22, 1972, pp. 1080–1082.

————, "Reactor Safety: AEC Concedes Some Points to Its Critics," *Science*, Nov. 3, 1972, pp. 482–484.

Gofman, J. W., and A. R. Tamplin, *Poisoned Power: The Case Against Nuclear Power Plants*, Rodale Press, 1971.

Gough, W. C., and B. J. Eastlund, "The Prospects of Fusion Power," *Scientific American*, Feb. 1971, pp. 50–64.

Hammond, A., "Fission: The Pro's and Con's of Nuclear Power," *Science*, Oct. 13, 1972, pp. 147–149.

Hammond, R. P., "Nuclear Power Risks," *American Scientist*, March/April, 1974, pp. 155–160.

Joint Committee on Atomic Energy, "The Safety of Nuclear Power Reactors (Light Water Cooled) and Related Facilities," Government Printing Office, Washington, D.C., Dec. 1972.

Lapp, R. E., "How Safe Are Nuclear Power Plants?" *New Republic*, Jan. 23, 1971, p. 19.

Leeper, C. K., "How Safe Are Reactor Emergency Cooling Systems?" *Physics Today*, Aug. 1973, pp. 30–35.

Metz, W. D., "Laser Fusion: A New Approach to Thermonuclear Power," *Science*, Sept. 29, 1972, pp. 1180–1182.

————, "Magnetic Containment Fusion: What Are the Prospects," *Science*, Oct. 20, 1972, pp. 291–293.

————, "Uranium Enrichment: U.S. 'One Ups' European Centrifuge Effort," *Science*, March 29, 1974, pp. 1270–1272.

Post, R. F., "Prospects for Fusion Power," *Physics Today*, April 1973, p. 31.

"Quicker Start of Nuclear Power Plants Sought," *Chemical and Engineering News*, Nov. 26, 1973, pp. 7–8.

Rose, D. J., "Controlled Nuclear Fusion: Status and Outlook," *Science*, May 21, 1971, pp. 797–808.

————, "Nuclear Electric Power," *Science*, April 19, 1974, pp. 351–359.

Sagan, L. A., "Human Costs of Nuclear Power," *Science*, Aug. 11, 1972, pp. 487–493.

Seaborg, G. T., and W. A. Corliss, *Man and Atom: Building a New World Through Nuclear Technology*, E. P. Dutton & Company, Inc., New York, 1971 (Chapter 2).

U.S. Atomic Energy Commission, "The Nuclear Industry — 1973," Publication WASH 1174–73, Government Printing Office, Washington, D.C., 1974.

Weinberg, A. M., "Social Institutions and Nuclear Energy," *Science*, July 7, 1972, pp. 27–34.

"Who Will Produce Enriched Uranium," *Chemical and Engineering News*, Dec. 3, 1973, pp. 8–9.

Chapter 9

Additional Energy Sources and Improved Energy Utilization

The preceding four chapters have dealt with the widely used fossil fuels and nuclear energy sources, considered by some to be the energy sources of the future. In this chapter we conclude our discussion of energy sources with a look at hydroelectric energy, a traditional source with a limited potential for expansion, and some nontraditional sources that have been gaining varying amounts of support as possible (partial or total) solutions to the energy crisis. Included in this latter category are geothermal, tidal, solar, wind, and ocean thermal gradient energy sources. We also include a discussion of some possibilities for improving energy utilization and conversion, including the use of secondary chemical fuels (hydrogen), fuel cells, and magnetohydrodynamics.

Hydroelectric Energy

Falling water from natural or regulated streamflow has been used as a source of useful energy for many centuries. The waterwheel, initially used in the first century B.C., was the first major source of energy that was not related to human or animal muscle power. By the sixteenth century, the waterwheel was western Europe's most important energy source. Steam engines, the first really mobile energy sources, had replaced most waterwheels by the middle of the nineteenth century. However, interest in water power was renewed during the 1850–1900 period as a result of the invention of the versatile and efficient hydraulic turbine. This device had a much greater maximum power output than waterwheels (see Figure 1–1 for the comparison). Numerous types of improved turbines have been developed since the first one dedicated to generating electricity was installed on the Fox River in Wisconsin in 1882. Today, the generation of hydroelectricity is common in many parts of the United States.

Hydroelectric Energy Use and Generation

The gross consumption of hydroelectric energy in the United States during 1973 was 2932 trillion Btu, an amount equal to 3.9% of the total energy used during the year. It was more than double that used 25 years previously, as shown by Figure 9—1. Despite the doubling of hydroelectric energy used in the period, the percentage contribution to total energy used has remained nearly constant at about 4%. This indicates that hydroelectric generating capacity has grown at about the same rate as the total U.S. energy demand. The contribution of hydroelectricity to total U.S. needs prior to 1947 can be seen in Figure 1—3.

The total hydroelectric generating capacity is projected to double again in the next 25 years. These projections are shown in Table 9—1. The hydroelectric growth rate during the projected period will be less than the growth rate of total energy demand, as indicated by the declining percentage contribution of hydropower to total needs.

Two types of systems are used to generate hydroelectric energy. In conventional systems, water passes through turbines while falling from a higher to a lower elevation that results from natural topography (waterfalls, etc.) or structures such as dams. In pumped storage systems, water is pumped uphill and stored in reservoirs until energy is needed, at which time it is discharged through turbines.

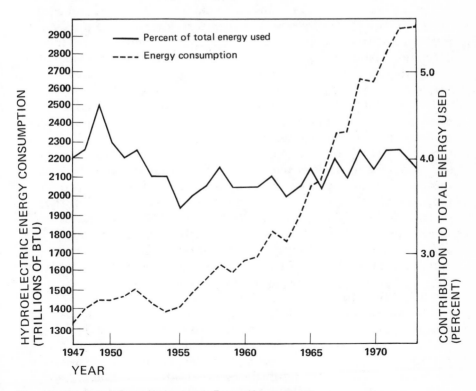

Figure 9—1.   United States Hydroelectric Energy Use, 1947—73

Data from "U.S. Energy Through the Year 2000," U.S. Dept. of the Interior, Dec. 1972, pp. 16, 40; "The President's Energy Message and Senate Bill 1570," U.S. Senate Subcommittee Hearings, May 1973, p. 35; and 1973 data from U.S. Dept. of the Interior, quoted in *Oil and Gas Journal*, March 25, 1974, p. 44.

Table 9—1.   PROJECTED GROWTH IN HYDROELECTRIC GENERATING CAPACITY

Year	Projected Generating Capacity (trillions of Btu)	Contribution to Total U.S. Demand (percent)
1972 (actual)	2937	4.1
1975	3570	4.4
1980	3990	4.2
1985	4320	3.7
2000	5950	3.1

Data from "U.S. Energy Through the Year 2000," U.S. Dept. of the Interior, Dec. 1972, p. 17.

Pumped storage systems are usually used to provide electricity during periods of peak demand on electrical distribution systems. During off-peak times, excess electricity generated by other sources is used to pump water back into the storage reservoir. More energy is required to store the water than can be recovered by releasing it again, but an economic advantage results. Low cost, low value off-peak energy is stored and later changed to high value peak demand energy. Thus, a generating system can be built to generate less than the peak amount of energy required of it. Pumped storage systems can be combined with fossil-fueled or nuclear-generating facilities; in this case the pumped water is used repeatedly and just moves between a high and a low reservoir. Pumped storage may also be combined with a conventional hydroelectric system, as illustrated in Figure 9—2.

The problem of satisfying peak energy demands is faced by all energy-generating utilities and is shown graphically in Figure 9—3. In this figure the hourly power demands of a large utility are shown. The wide fluctuations during a day are quite obvious. About 50% of the weekly maximum demand is essentially constant and is called the base demand or load. The intermediate load amounts to approximately 30% of the maximum and is continuous for periods of 12 or more hours during weekdays. The peak portion of the total load amounts to about 20% of the maximum and occurs over periods of less than 1 hour to about 12 hours on weekdays.

The availability of suitable sites for pumped storage systems depends on the topography of the region; it must allow for the construction of reservoirs with large elevation differences between them. Storage reservoir capacities must be large enough to assure dependable operation even under the most adverse conditions. The load pattern shown in Figure 9—3 would only allow water to be pumped and stored for about 10 hours each weekday night. Normally, the pumping cycle requires 50% more time than the outflow cycle, so a daily 10-hour pumping cycle would only assure generating time of 6 or 7 hours per day. In some cases this would not satisfy the peak demand period. For this reason, it is often necessary to construct storage reservoirs with sufficient capacity to supply a weekly cycle which includes substantial pumping on weekends when peak demands are low.

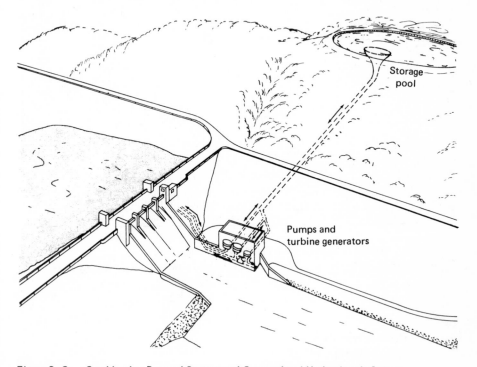

Storage pool

Pumps and turbine generators

Figure 9—2.   Combination Pumped Storage and Conventional Hydroelectric System

From "The 1970 National Power Survey," Part 1, Federal Power Commission, Dec. 1971, p. I—7—7.

At the end of 1970, conventional hydroelectric capacity totaled 52,323 megawatts. This compares to only 3689 mw of pumped storage capacity. However, the installation of pumped storage generating capacity is expected to greatly exceed that of conventional capacity during the next 15 years. By 1990, it is estimated that pumped storage hydroelectric capacity will be almost equal to conventional capacity as shown by the data of Figure 9—4. Pumped storage capacity amounted to only 7% of the total hydroelectric generating ability in 1970 but is expected to increase to 28% by 1980 and to 46% by 1990.

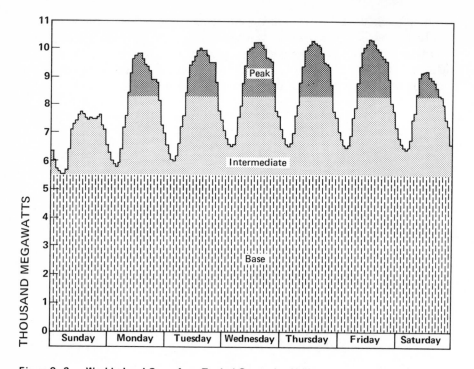

Figure 9–3.   Weekly Load Curve for a Typical Generating Utility

From "The 1970 National Power Survey," Part 1, Federal Power Commission, Dec. 1971, p. I–3–2.

Hydroelectric Versus Thermal Plants

Hydroelectric generating plants have some distinct advantages when compared to thermal (steam turbine) types. These advantages include the following:

1.   Hydroelectric plants have lower operational and maintenance costs.
2.   Hydroelectric plants are more easily designed to operate automatically or under control from remote locations.
3.   No fuel is required to operate hydroelectric plants except that needed to run pumps in some pumped storage systems.
4.   Hydroelectric plants have long lives and depreciate slowly.
5.   Hydroelectric plants are well adapted to quick starts and rapid changes in output. Thus, they are very useful in satisfying peak demands.
6.   No polluting emissions are released into the atmosphere and no heat is discharged into receiving waters.

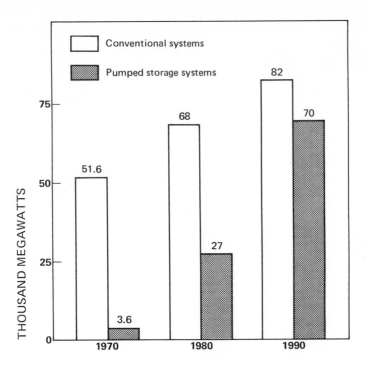

Figure 9—4.   Comparison of Conventional and Pumped Storage Hydroelectric Generating
Capacities

From "Final Environmental Statement for the Geothermal Leasing Program," vol. 1, U.S. Dept. of the Interior, 1973, p. IV—173.

Location of Conventional Hydroelectric Facilities

The potential generating capacity of any conventional hydro-electric plant depends on the volume of water available for use, the regulation of water flow, and the height the water falls. Consequently, the best hydroelectric sites are found in areas of heavy precipitation that are also characterized by significant variations in elevation. Figure 9—5 shows the location of the principal hydroelectric projects (25-megawatt capacity or greater) in operation or under construction in the United States. It is apparent that the projects are concentrated in two areas, the Northwest and the Southeast. The total capacity located in the Northwest is much larger than is found in any other geographical area. This fact is reflected by the data of Figure 9—6, where the geographical regions are the same as those shown earlier in Figure 6—7.

One third of the total U.S. hydroelectric capacity is located within the Columbia river basin. Six of the ten largest projects are found in this area, and all but one of the six are located in the state of Washing-

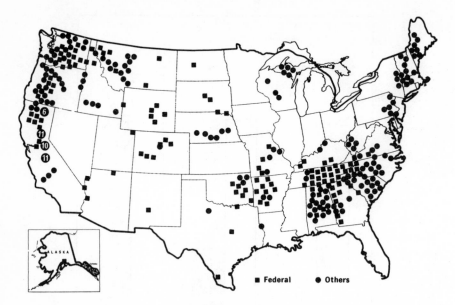

Figure 9—5.   Locations of the Principal Hydroelectric Projects in the United States
From "River of Life," Environmental Report — 1970, U.S. Dept. of the Interior, p. 33.

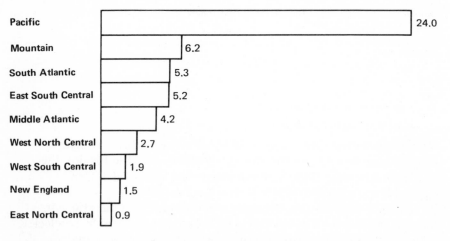

Figure 9—6.   Developed U.S. Hydroelectric Capacity, December 31, 1970 (thousands of megawatts)

Data selected from "Final Environmental Statement for the Geothermal Leasing Program," vol. 1, U.S. Dept. of the Interior, 1973, p. IV—170.

ton. The output of the largest project is more than 5000 megawatts and six others are rated at more than 1000 megawatts. The ten largest projects are listed in Table 9—2. Those found in the Columbia river basin are indicated by an asterisk.

Table 9–2.   THE TEN LARGEST U.S. HYDROELECTRIC PROJECTS
OPERATING OR UNDER CONSTRUCTION
DECEMBER 31, 1970

| | Installed Output (megawatts) | | |
Project Location	Existing	Under Construction	Total
*Grand Coulee, Washington	2066	3600	5666
*John Day, Washington	1890	270	2160
Niagara, New York	1954	–	1954
*The Dalles, Washington	1119	688	1807
Hoover, Arizona-New Mexico	1340	–	1340
*Rocky Reach, Washington	712	502	1214
*Chief Joseph, Washington	1024	–	1024
*McNary, Oregon	980	–	980
Glen Canyon, Arizona	950	–	950
Robert Moses, New York	912	–	912

Data selected from "The 1970 National Power Survey," Part 1, Federal
Power Commission, Dec. 1971, Table 7.4, pp. I–7–13 to I–7–16.

Future Hydroelectric Development

The total conventional hydroelectric potential in the United States
is estimated to be 179,900 megawatts. Of this amount, 51,900 megawatts or 28.8% is already developed or under construction. The
remaining 128,000 megawatts of capacity includes projects for which
the economic and engineering feasibilities have been proven as well as
projects with suggested engineering feasibility but which lack detailed
economic information. The amount of hydroelectric potential, both
developed and undeveloped, contained in the various regions of the
United States is given in Table 9–3.

The development of this large amount of undeveloped generating
capacity is limited by a number of factors. Alaska contains 36,600 of
the 128,000-megawatt potential. The state's sparse population and the
distance of the potential sites from population centers casts doubt on
the economic feasibility of development. Economics may curtail
development of other sites as well. Costs for installing conventional
hydroelectric facilities depend upon such things as the project size, land
acquisition costs, and the expense of relocating disrupted facilities
(railroads, bridges, roads, etc.). The cost variations are large from one
project to another, but on the average, investment costs per kilowatt
are higher for hydroelectric plants than for nuclear or fossil-fueled
projects. Increasing fuel costs could make hydroelectric power more

Table 9–3.   HYDROELECTRIC POTENTIAL OF THE UNITED STATES

Region	Conventional Capacity (thousands of megawatts)			
	Total Potential	Developed Capacity	Undeveloped Capacity	Percent Developed
New England	4.8	1.5	3.3	31.3
Middle Atlantic	8.7	4.2	4.5	48.3
East North Central	2.5	0.9	1.6	36.0
West North Central	7.1	2.7	4.4	38.0
South Atlantic	14.8	5.3	9.5	35.8
East South Central	9.0	5.2	3.8	57.8
West South Central	5.2	1.9	3.3	36.5
Mountain	32.9	6.2	26.7	18.8
Pacific	62.2	23.9	38.3	38.4
Alaska	32.6	0.1	32.5	0.3
Hawaii	0.1	—	—	—
Total	179.9	51.9	128.0	

Data selected from "Final Environmental Statement for the Geothermal Leasing Program," vol. 1, U.S. Dept. of the Interior, 1973, p. IV–170.

economically attractive in the future. However, another limiting factor is the prohibition on developing some sites. The Colorado River Basin Project Act, for example, prohibits the Federal Power Commission from issuing licenses for any power development on the Colorado River between the Glen Canyon and Hoover Dam projects — a part of the Colorado River with a potential for generating 3500 megawatts. Similarly, the Wild and Scenic Rivers Act prevents the FPC from licensing the construction of power facilities that would affect any river included in the national and scenic rivers system established by the act. Thirty-seven rivers could be involved with a total hydroelectric potential of about 9000 megawatts.

Environmental Effects of Hydropower Development

Conventional hydroelectric facility developments have several beneficial effects. The resulting reservoirs are used for recreation and as water supplies for domestic and industrial use. Some projects provide flood control during wet seasons and irrigation or other water during dry seasons.

However, some adverse effects are also associated with hydroelectric installations. Below the dams the water level is lowered, and above the dams land areas are submerged beneath lakes. Some of the resources lost include arable land, wildlife habitats, minerals, timber areas, historical and archeological sites, and free-flowing river recreation. Fish and wildlife habitats can be dramatically affected. The reproductive habits of up-river migrating fish such as salmon may be severely altered or curtailed by dam construction unless appropriate measures are taken — fish ladders, etc.

The mortality rate of resident fish below a dam can be increased by nitrogen narcosis (gas-bubble disease) resulting from exposure to water supersaturated with nitrogen. Nitrogen supersaturation is caused by excessive mixing of air with water during its passage over the spillways of dams. Research aimed at solving this problem is in progress.

Geothermal Energy

As the name indicates, geothermal energy is made up of heat from the earth. The temperature inside the earth increases with increasing depth; at depths of 15—30 miles (at the base of the earth's crust), temperatures between 200° and 1000°C (400°—1950°F) are found. The temperature at the center of the earth has been estimated to be 3500°—4500°C (7300°—8100°F). Most of this vast amount of energy is located too far under the surface to be recovered and used. The deepest drilling into the earth has been to a depth of only about 4.7 miles. It is thought that this might be extended to depths as great as 10—12 miles, but the depth from which heat might be profitably extracted is probably no greater than about 6 miles. At this depth, most of the heat is quite diffuse and cannot be recovered economically.

In some locations, however, there are significant concentrations of geothermal energy. In such "hot spots," temperatures of 65°—343°C (150°—650°F) are found at depths of less than 2 miles. Such concentrations of heat are analogous to concentrations of minerals in ore deposits or petroleum in reservoirs. These geothermal reservoirs are usually found in regions characterized by recent (on the geologic time scale) volcanic and mountain-building activity. Most of the easily accessible geothermal deposits in the United States are believed to be in the West, although some eastern hot springs have been explored with a view toward using them for energy production. Figure 9—7 shows the location of known and likely western geothermal areas.

Geothermal reservoirs result when magma, a mixture of molten rock and gases, penetrates into the earth's crust from the underlying mantle. When the magma penetrates through to the surface, it erupts in a volcano. But when the penetration stops short of the surface, the trapped magma heats rocks near the surface and local hot spots, or geothermal reservoirs, are created. The resulting energy sometimes remains in the form of hot rocks, or it may be transferred to underground water and form hot water or steam. Geothermal reservoirs are

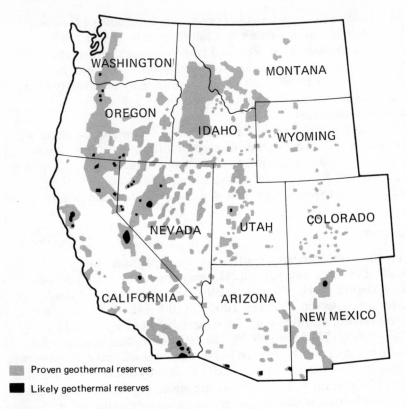

Figure 9—7.   Geothermal Areas of the Western United States

Redrawn with permission from "Geothermal Heats Up," *Environmental Science and Technology*, 7 (August 1973): 680. Copyright by American Chemical Society.

classified into one of the following categories on the basis of the form in which the energy is found: (1) vapor-dominated systems, (2) hot water systems, and (3) hot, dry rock systems. Figure 9—8 illustrates the formation of these various types of geothermal reservoirs.

Vapor-dominated Systems

The least common and easiest geothermal systems to work with are those classified as vapor-dominated. In this type of system, the energy is found primarily in the form of superheated steam that contains minor amounts of other gases (CO_2, H_2S, NH_3) but little or no liquid water. For this reason, they are sometimes called *dry steam systems*. The energy of such systems is tapped by drilling a well into the reservoir and allowing the superheated and pressurized steam to flow through appropriate pipes to the point of use. The steam is usually filtered to remove abrasive particles and then passed through turbines

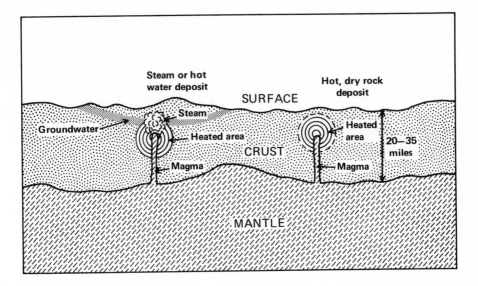

Figure 9—8.   The Formation of Geothermal Energy Reservoirs

to generate electricity. Most of the steam and condensed water from the turbines is released into the surrounding environment. However, condensed water which contains polluting trace minerals (about 20% of the total) is injected back underground by way of deep wells.

Since vapor-dominated systems are the easiest to work with, they were the first to be commercialized. The first successful operation began in 1904 at Larderello, Italy. The Larderello field is still producing electricity and has a current output of 365 megawatts. The only commercial use of geothermal energy in the United States began in 1960 when plants located at The Geysers steam field in Sonoma County, California, went into production. This operation, located 80 miles north of San Francisco, began with one generator and an output of 12.5 megawatts. By the end of 1973, ten generators were in use and the output had risen to slightly more than 400 megawatts — the largest output of any geothermal operation in the world. It is anticipated that 110 megawatts of output will be added annually during the 1970s so that by 1980, an output of about 1180 megawatts will be achieved — more than enough to supply all of the electricity demands for a city the size of San Francisco.

Current geothermal electricity-generating plants differ somewhat from fossil-fueled plants. Geothermal plants require more steam per unit of electricity produced. They are much smaller than fossil-fueled plants, partially because no fuel is combusted in a geothermal plant.

More steam is required by geothermal plants because of the comparatively low pressure and temperature at which the steam reaches the turbines — 100 pounds per square inch and 205° C (400° F). Turbines used at The Geysers steam field are designed differently and are about one-third less efficient than those of a conventional plant. In spite of the lower efficiency, the geothermal plants are smaller, cost less to build, and are cheaper to operate than either fossil-fueled or nuclear power plants. No combustion furnaces, fuel stockpiles, boilers, or radioactive shielding are required for the geothermal plants. Table 9—4 contains estimates of the cost of electricity generated using various energy sources in the western United States.

Table 9—4.   COMPARISON OF ENERGY COSTS ACCORDING TO POWER PLANT

Power Plant Type	Energy Costs ($/mwhr)
Geothermal (vapor-dominated)	5.25
Gas fired (steam)	6.70
Coal fired (steam)	8.20
Hydroelectric	9.60
Nuclear (steam)	9.60
Geothermal (hot water)	9.70
Oil fired (steam)	10.00

From "Final Environmental Statement for the Geothermal Leasing Program," vol. 1, U.S. Dept. of the Interior, 1973, p. II—19.

Hot Water Systems

Water, confined under pressure and superheated to temperatures well above the normal boiling point (180°—370° C or 350°—700° F), constitutes the energy potential of hot water systems. On the basis of discoveries made to date, it appears that hot water systems may be 20 times more abundant than vapor-dominated systems. When the superheated water is brought to the surface and exposed to normal atmospheric pressure, it boils rapidly (flashes), and a mixture of steam and hot water results. Only about 10—20% of the weight of superheated water is converted to steam in the process. The large amount of hot water found with the steam has led to the practice of calling such deposits *wet steam reservoirs.* Only the steam from these systems can be utilized with present technology, so it is necessary to separate the steam from the hot water (really a brine of mineral salts dissolved in water) at the wellhead. The steam is then passed through a turbine to generate electricity.

Proven United States hot water systems are located in California, Nevada, New Mexico, Oregon, and Idaho, and potential but unproven sites have been identified in other western states. None of these domestic deposits have yet been commercialized, but there are some commercial hot water plants producing electricity in other countries. For example, a plant currently generating 192 megawatts has been operating for nearly 20 years in Wairakei, New Zealand, and a 75-megawatt plant located just south of the U.S. border near Cerro Prieto, Mexico, began operating in 1973.

A great deal of effort is being directed toward developing processes which will enable some of the energy contained in the hot brines of wet steam systems to be utilized. The processes under study involve heat exchangers and low-boiling-point secondary fluids. Heat from the hot brines is transferred to the secondary fluid by the heat exchanger. The secondary fluid is vaporized, and the resulting vapors are used to drive a turbine. After expanding through the turbine, the secondary fluid is condensed to liquid and returned to the heat exchanger, where the cycle begins again. The geothermal brines, after passing through the heat exchanger, can be reinjected into the geothermal reservoir.

The costs for power plants based on secondary fluid technology are uncertain and vary with the temperature of the available brines. Under favorable conditions the price of electricity from such plants is estimated to be competitive with that from fossil-fueled plants.

A plant using secondary fluid technology is now operating in the U.S.S.R., and a 10-megawatt prototype unit is scheduled to be built near the Salton Sea in California. Figure 9—9 shows a possible flow diagram for a geothermal plant utilizing a secondary fluid power cycle. The pump located near the producing zone in the well is used to keep the brine under pressure and prevent it from flashing into a steam and hot water mixture. In some cases, the brine may be treated and solids removed to prevent the reinjected brine from clogging the pores of the reservoir. In other cases, the brine may be treated to extract valuable mineral by-products. The successful development of secondary fluid technology will make it possible to extract more of the energy found in the higher temperature hot water systems and will make possible the generation of electricity from reservoirs with temperatures too low for steam production — a situation found in a large number of geothermal deposits.

The possibility of using geothermal water for other than energy production is being investigated. Where water is in short supply, hot water power plants might be used as sources of fresh water. The water from the geothermal wells is already hot, and desalting techniques similar to those being developed for seawater may prove to be relatively inexpensive. The condensed turbine exhaust might also be used as a source of fresh water. Figure 9—10 contains a flow chart for a combined power-generating and water desalinization operation. This

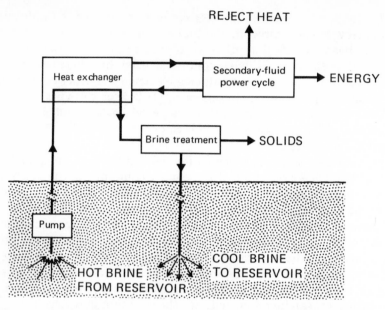

Figure 9—9.   Geothermal Energy Extraction Using a Secondary-Fluid Cycle

Redrawn with permission from "Water from Geothermal Resources" by A. D. K. Laird, p. 181, in *Geothermal Energy: Resources, Production, Stimulation*, P. Kruger and C. Otte, eds., Stanford University Press, 1973.

operation also allows for the recovery of valuable minerals from the brine. A system such as this is being designed for a new geothermal field in Chile as a joint effort of the Chilean government and the United Nations.

Hot, Dry Rock Systems

Most geothermal heat deposits located near the earth's surface never come in contact with underground water and therefore do not contain steam or hot water. These dry thermal deposits, or hot, dry rock systems, are believed to constitute an energy resource at least 10 times larger than the total of all vapor-dominated and hot water systems. The techniques needed to extract energy economically from hot, dry rock deposits apparently exist, but they have not yet been applied to such ends.

Preliminary analyses indicate that it should be possible to recover energy from dry geothermal deposits by injecting cool water and extracting the resulting hot water. The process would be accomplished in several steps. First, a hole would be drilled into a suitable deposit having temperatures of at least 300°C (572°F). Pressurized water injected through this hole would be used to produce large cracks in the deposit (hydrofracturing). The cracks provide a large area for the

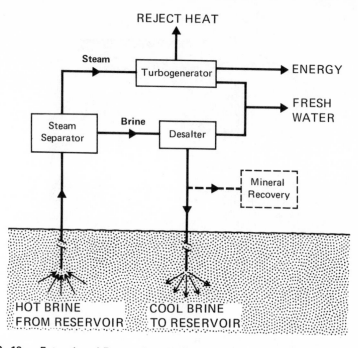

REJECT HEAT

Steam

Turbogenerator → ENERGY

FRESH
WATER

Steam
Separator

Brine

Desalter

Mineral
Recovery

HOT BRINE
FROM RESERVOIR

COOL BRINE
TO RESERVOIR

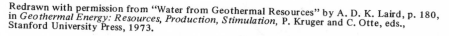

**Figure 9—10.   Extraction of Energy, Fresh Water, and Minerals from Geothermal Hot Water
Deposits**

Redrawn with permission from "Water from Geothermal Resources" by A. D. K. Laird, p. 180,
in *Geothermal Energy: Resources, Production, Stimulation,* P. Kruger and C. Otte, eds.,
Stanford University Press, 1973.

transfer of heat from the deposit to water introduced through the hole.
A second hole drilled to intersect the deposit above the original hole
would be used to remove the heated water. The entire system would be
kept under pressure to prevent the water from boiling. Hot water
pumped to the surface would be used in a secondary fluid cycle to
generate electricity. The now-cooler water would be circulated back
into the heat deposit. Such a proposed system is shown in Figure 9—11.

The first field trials designed to test the feasibility of these
procedures were done in 1973 by the AEC's Los Alamos scientific
laboratory. The site was located on the edge of an extinct volcano near
Los Alamos, New Mexico. A hole was drilled one-half mile deep into a
rock formation. Pressurized water was then injected, and it successfully
fractured the rock formation. The results, in addition to proving the
feasibility of hydraulically fracturing granite and other crystalline
rocks, indicated that the fractured rock systems allowed very little water
to leak away — a necessary condition for the success of the proposed
heat-recovery techniques. The next experiment planned by the Los
Alamos group will involve the drilling of an experimental well about one
mile deep.

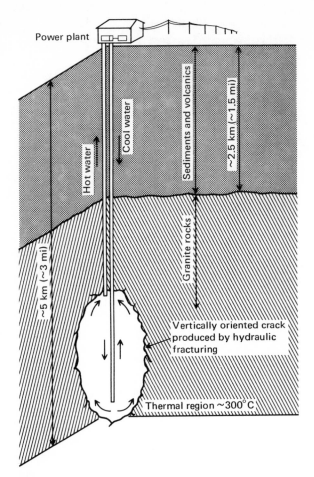

Figure 9—11.   Extraction of Energy from Hot, Dry Rock Deposits

Redrawn with permission from "Dry Geothermal Wells: Promising Experimental Results" by A. L. Hammond, *Science*, 182 (Oct. 5, 1973): Fig. 1, p. 43. Copyright 1973 by the American Association for the Advancement of Science.

Environmental Impact of Geothermal Development

Geothermal energy has received considerable attention as an example of an energy source that has little adverse effect on the environment. The facts often used are:

1. Geothermal energy, unlike fossil-fueled plants, produces no particulate air pollutants.
2. Geothermal energy does not require the mining, transporting, handling, or consumption of any fuels.
3. Geothermal energy generates no radioactive wastes such as those from fission reactor power plants.
4. No supplemental cooling water is needed to run geothermal plants.

Despite these positive features, the development and use of geothermal energy can adversely influence the environment. Among the potential problems are:

1. A typical geothermal field and power plant require more land area than an equivalent nuclear or fossil-fueled plant. Currently, a geothermal development makes use of several wells, each one located on a 20-acre site, to supply a small central power plant. The involved area is laced with pipes radiating from power plants which in turn are connected by high-voltage power lines.

2. Small amounts of noncondensable and sometimes objectionable gases are associated with the steam obtained from geothermal sources. Table 9—5 shows the gases found in the steam of the two largest commercial geothermal sources. Hydrogen sulfide (H_2S) is the most objectionable component because of its unpleasant odor and its toxicity. The odor can be detected when the gas is in concentrations as low as 0.025 parts per million. The undiluted values given in the table are 330 and 490 ppm. Normally, H_2S would mix with the atmosphere, become diluted, and pose little problem. However, during temperature inversions or times of little air movement, H_2S could accumulate to nuisance and possibly toxic levels in the region around a power plant. At the present time no completely satisfactory technique exists to remove it from steam. Research is continuing in this area, but further restrictions on sulfur emissions into the atmosphere might provide more incentives to use secondary fluid systems in which atmospheric emissions are more easily controlled.

Table 9—5.   **GASES ASSOCIATED WITH GEOTHERMAL STEAM**
(volume percent)

	The Geysers, California	Larderello, Italy
H_2O	98.045	98.08
CO_2	1.242	1.786
H_2	0.287	0.037
CH_4	0.299	
N_2	0.069	0.0105
H_2S	0.033	0.049
NH_3	0.025	0.033
H_3PO_4	0.0018	0.0075

Data selected from "Final Environmental Statement for the Geothermal Leasing Program," vol. 1, U.S. Dept. of the Interior, 1973, p. III—13.

3. The disposal of waste waters from steam of hot water wells could present chemical water pollution problems, especially when the waste water is highly mineralized. The dissolved salt content of geothermal waters from an area near the California Salton Sea is as high as 20%. Seawater, by way of contrast, contains about 3.3% dissolved salts. Waters with such high salt content will have to be injected back into deep wells or treated to remove salts before being released into the environment.

4. The possibility of land subsidence (settling) exists when large amounts of water or other liquids are removed from beneath the ground. This has taken place in some types of oil fields and has been alleviated to some extent by injecting water into the wells after removing the oil.

5. Nearly all of the cooling methods being considered for geothermal plants release waste heat into the atmosphere. The atmosphere is generally able to absorb heat with less environmental impact than that exerted on natural waters. However, the effects of such heat on local weather patterns is not yet known.

Problems in the Development of Geothermal Resources

The projected contribution of geothermal-based electricity to the total amount of generated electricity during the next quarter of a century is given in Table 9—6. Two different estimates are given, based upon different research and development programs. The moderate research and development program estimates include assumptions that technologies will be developed for the economic generation of electrical power from hot water reservoirs. An accelerated program could bring about an earlier application of these technologies and result in a much

Table 9–6.   CONTRIBUTION OF GEOTHERMAL ELECTRICITY TO TOTAL U.S. OUTPUT (megawatts of power)

Year	Results of a Moderate R and D Program	Results of an Intensive, Accelerated R and D Program	Total Generating Capacity from All Sources
1980	10,500	36,000	660,000
1985	19,000	132,000	915,000
2000	75,000	395,000	1,880,000

Data from "Geothermal Energy Resources and Research," U.S. Senate Subcommittee Hearings, June 1972, p. 245 and "U.S. Energy Through the Year 2000," U.S. Dept. of the Interior, Dec. 1972, p. 19.

greater capacity by 1985. The accelerated estimate also assumes the development of processes for extracting heat from hot, dry rock sources. The optimistic figures were obtained by also assuming complete success in solving all other principal technological problems related to geothermal energy development. Thus, optimistic figures represent the maximum amount of geothermal energy available under the most favorable circumstances.

The significance of the geothermal energy potential can be seen by comparing the estimated geothermal output to the estimated total United States generating capacity (see Table 9—7). It is apparent that geothermal electricity, even under the most favorable circumstances, is not going to replace that from fossil-fueled or nuclear sources. However, it could satisfy a small but important segment of the market.

Problems other than technology could exert strong influences on the development of geothermal energy sources. One of these is the federal ownership of land. About 1.8 million acres of western land have been classified by the U.S. Geological Survey as being within Known Geothermal Resource Areas (KGRA). An additional 96 million acres are listed as having prospective value for geothermal resources. About one million of the 1.8 million acres of KGRA land is federally owned. Federal ownership is also involved in approximately 58 million of the 96 million acres of land with prospective value. Thus the federal government owns about 56% of the prime and 60% of the potential geothermal land. The Geothermal Steam Act of 1970 provides for the leasing of some of this land for purposes of geothermal resource development. The frequency and extent of lease granting will have a bearing on development patterns.

Another problem is one related to the lifetime of geothermal reservoirs once they are placed into production. This useful lifetime is not known, and the uncertainty may be a factor which will retard investments in the commercial development of geothermal sources. The Internal Revenue Service recognized the problem and has already

Table 9—7.   PERCENTAGE CONTRIBUTION OF GEOTHERMAL ELECTRICITY TO TOTAL U.S. OUTPUT

	Moderate R & D Program	Intensive R & D Program
1980	1.6	5.5
1985	2.1	14.4
2000	4.0	21.0

Calculated from data in Table 9-6.

decided that geothermal steam, like petroleum and natural gas, is a depletable resource. Little experience is available on which to base estimated useful lives. It is not even known whether steam or hot water is replaced naturally in geothermal wells. Production and pressure data from The Geysers field in California indicate a decline in production with time due to pressure depletion. Figure 9—12 shows data collected for two wells of this field. These data seem to indicate a depletion of geothermal steam is taking place, since other tests show no evidence of physical changes in the well bore (plugging of pores, etc.) that would cause the observed declining output. Geothermal reservoirs might be stimulated by techniques similar to those used in the gas and petroleum industry. Studies in this area are in progress.

Tidal Energy

The tidal movement of ocean water represents a large amount of kinetic energy that can, in principle, be converted into other, more useful forms. Most coastal areas of the earth experience two tides daily — twice the water level rises and twice it recedes. Various ways of

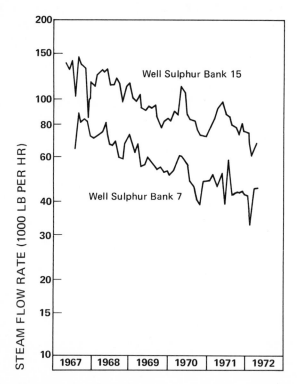

Figure 9—12.   Examples of Declining Geothermal Steam Output with Time

Redrawn with permission from "Steam Production at The Geysers Geothermal Field" by C. F. Budd, Jr., p. 136, in *Geothermal Energy: Resources, Production, Stimulation*, P. Kruger and C. Otte, eds., Stanford University Press, 1973.

harnessing some of this energy are being studied and in at least two locations are now in use. The feasibility of extracting energy from the tides is dependent on the tidal behavior and range (difference in level between low and high tide) and on the amount of water displaced.

Origin and Variation of Tides

The tidal behavior characteristic of the world's oceans and seas results from the action of gravitational and centrifugal forces on water. The gravitational forces are exerted on the earth by the sun and moon. The centrifugal forces are the result of the revolutions of the earth and moon about the sun. The combination of these two forces causes water to simultaneously accumulate on the side of the earth nearest the moon and on the side directly opposite. As the earth rotates relative to the moon, these bulges of water are forced against the shore and high tides result. Since there are two bulges, two high tides (and two low tides) occur during each daily rotation. The sun and the moon both influence the tides. Since their relative positions and hence their effect on the tides is not always the same, tidal variations occur. When the sun, moon, and earth are approximately in line, the forces combine to produce extremely high and low tides. When the sun and moon form a right angle with the earth, the effects tend to cancel one another and high and low water are not so extreme. Figure 9—13 shows the effects of these two arrangements, with the effect on tidal height indicated by the black band around the earth.

Other factors also influence the tidal flow of water, including the shape of the ocean floor and the latitude of the shoreline. Tidal ranges are lowest near the equator and highest near the poles. The result of the interaction of all factors is a great variation in tidal behavior in different parts of the world.

Along the United States east coast, the tidal range changes from 18 feet at Eastport, Maine, to 9 feet at Cape Cod, Massachusetts. Moving south from Cape Cod, the range decreases to about 2 feet at Florida. On the Gulf Coast the range is generally less than 2 feet. On the west coast, it changes from 4 feet at San Diego to slightly more than 11 feet at Seattle. Alaskan and Canadian tides have ranges between 12 and 18 feet.

Operation of a Tidal Power Plant

Tidal energy is extracted by alternately filling and emptying a bay or estuary that can be closed by a dam. During a part of both the filling and emptying cycles, the water is diverted through electricity-generating turbines. The conditions necessary for a successful operation include an adequate tidal range and geographical features that allow a basin to be created by a dam.

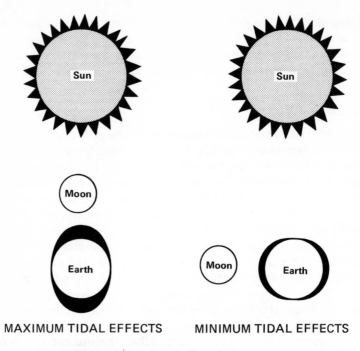

MAXIMUM TIDAL EFFECTS   MINIMUM TIDAL EFFECTS

Figure 9—13.   Effect of Sun and Moon Orientation on Tides

Adapted from *Botany: An Ecological Approach* by William A. Jensen and Frank B. Salisbury, p. 303, © 1972 by Wadsworth Publishing Company, Inc., Belmont, California 94002. Reprinted by permission of the publisher, courtesy of Robert F. Scagel, et al.

Two commercial tidal electric installations are now in use, neither one in the United States. The largest of the two is located on the estuary of the Rance River near St. Malo, France. It was completed in 1967 at a cost of $90 million; its output is 240 megawatts. The second installation was completed in 1969 by the U.S.S.R. It is located on the coast of the White Sea and has an output of 1 megawatt.

The design of the turbines is a very important aspect of a tidal electric installation. They must be able to function with only a small head of water, and they must operate when the water flows through them in either direction. This latter characteristic is achieved by using moveable blades. The turbine generators used at the Rance River installation can also function as pumps when electricity is fed to them. The desirability of this feature will become apparent later.

Figure 9—14 illustrates the various operational phases of the Rance River facility when no water pumping is used. The sea level is shown by the solid line and the water level within the basin by the dashed line. The most important item to note is the vertical difference between these two lines which gives the height of the water head. When the head is large, power generation is possible. The general operation sequence is as follows:

1.   The gates are closed between the basin and sea until sea level increases and builds up a water head.

2.   The gates are opened and water flows from the sea, through the turbines, and into the basin. Electricity is generated.

3.   The turbines are stopped, but the gates are left open to allow further filling of the basin.

4.   The gates are closed when basin level and sea level are the same.

5.   During a waiting period, sea level falls and a new water head is created, in the opposite direction.

6.   The gates are opened and water flows out of the basin through the turbines — with blades reversed — and electricity is generated until the head becomes too small.

7.   The turbines are shut down, but the gates remain open to allow continuing outflow from the basin.

8.   When the basin and the sea are at the same level, the cycle begins again.

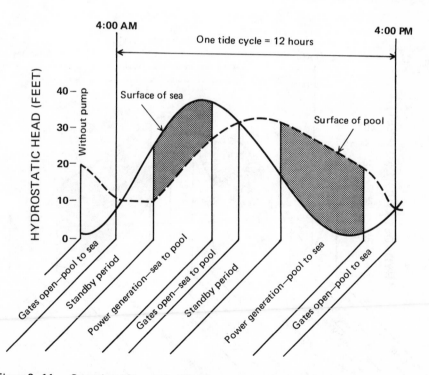

Figure 9—14.   Operational Phases of a Tidal Power Plant Without Water Pumping

From *The Tides* by E. P. Clancy. Copyright © 1968 by Doubleday & Company, Inc. Reproduced by permission of the publisher.

The use of the turbines as water pumps significantly increases the output of the Rance River facility. Pumping water into or out of the basin at appropriate stages of the operation increases the water head and the power production during generating stages. The annual non-pumped output at Rance River is 540 million kilowatt hours. Pumping increases this to 670 million kilowatt hours. Figure 9—15 shows how the water head is influenced by the pumping procedures.

The intermittent nature of the output from tidal power plants is their biggest disadvantage. The effect of this on-off operation can be seen by the following comparison. The Rance River plant has a maximum output of 240 megawatts. A conventional hydroelectric plant running with such an output would generate 2100 million kilowatt hours annually. The Rance River output is 670 million kilowatt hours annually or about one third as great. Consequently, tidal power development has been by-passed in favor of ordinary hydroelectric installations on rivers where dams are easier to build and a constant water head can usually be maintained. However, it should be mentioned that dry years sometimes affect hydroelectric plant output; there are no dry years for tidal plants.

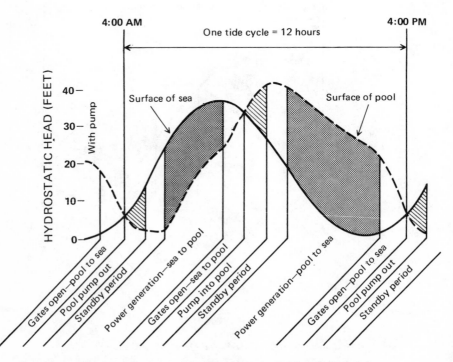

Figure 9—15.   Operational Phases of a Tidal Power Plant with Water Pumping

Tidal Power in the United States

Tidal ranges and geography suitable for harnessing tidal power with current technology exist only in the Northeast United States and Alaska. The Bay of Fundy between Maine and Canada contains nine suitable sites (some in Canada) with a total output capacity estimated at 29,000 megawatts. Turnagain Bay in the Cook Inlet of Alaska has a potential output capacity of 9500 megawatts, but its distance from major population centers makes it a doubtful economic risk. If the Bay of Fundy were developed and the output equally divided, the United States share of 15,000 megawatts would represent about 1.1% of the total electricity requirements by 1990. The project, if carried out, would cost approximately $1 billion.

The harnessing of tidal power cannot take place without affecting the environment. The damming and subsequent filling and emptying of bays and estuaries is likely to adversely affect some sport and commercial fishing, certain types of wildlife, water quality, recreational use of waters, shipping, and some aesthetic values.

Solar Energy

Solar energy arrives at the earth in the form of electromagnetic radiation covering a wide range of wavelengths and energies (see Figure 4—2). The quantities that reach the surface are sufficient to represent the greatest single potential energy source available on earth. The amount falling on just 0.5% of the United States land area is greater than the total domestic demand projected for the year 2000. However, it is so diffuse and intermittent in nature that its use hinges upon the development of suitable collection and storage systems. Despite growing interest, relatively little financial support and research have been devoted to the development of such systems. Some advocates of solar energy development maintain that appropriate support of research can lead to widespread commercial use of solar energy by the early 1980s. The uses of solar energy referred to are in addition to the natural uses evident in the form of agriculture and commercial forestry (photosynthesis), the hydrologic cycle, winds, and ocean currents. In other words, we are discussing applications that could replace other traditional energy supplies, primarily fossil fuels.

Extent of Solar Energy Use

The harnessing of solar energy has been a goal for hundreds of years, as shown by the brief (and incomplete) list of achievements given in Table 9—8.

Table 9—8.   PAST USES OF SOLAR ENERGY

Date	Solar Energy Use
1772	A solar furnace was built in Europe to melt metals.
1870	Solar distillation was used to provide fresh water to a remote mining area of Chile. The technique was used for 40 years.
Early 1900s	Solar water heaters were widely used in California and Florida. Some are still in use.
1901	A 4.5-horsepower solar-powered steam engine was developed in Pasadena, California.
1902—08	Solar engines capable of developing up to 20 horsepower were built in St. Louis, Missouri, and Needles, California.
1913	A 50-horsepower solar engine was built near Cairo, Egypt.
1946	Attempts were made to develop and distribute solar cooking stoves in India.
1948	A solar-heated house was operated at the Massachusetts Institute of Technology.
1959	A solar-heated laboratory was completed at the University of Arizona.
1961	Small vapor turbines using solar heat to generate electricity were demonstrated in Italy.

The use of solar energy is not totally dependent on the development of feasible systems. For example, the cooking stoves mentioned in Table 9—8 were developed in India for the 300 million peasant farmers. The stoves worked, but after one year of production, sales amounted to 50 units — in spite of a $14 selling price. The developers forgot that farmers usually eat their main meal at night. In addition, they overlooked the possibility that the wives would decline to cook outdoors in midday temperatures that often reached 100° F. Thus, we see that the behavioral patterns and habits of people must also be taken into account.

Economics must also be considered. According to some experts, a totally solar-powered home is feasible with current technology. It has a 99% probability for technological success, but the probability of achieving an economically competitive system under such conditions is no better than 50%. Solar-powered homes are currently so expensive that people won't buy them. As the costs of other energy sources increase and less expensive solar technology is developed, the economics will become more favorable.

The current use of solar energy is primarily limited to the following categories:

1.   Applications are made in which cost is relatively unimportant —
     The solar generation of electricity using solar cells is one of a very
     limited number of practical methods available for spacecraft. It is
     therefore used in spite of costs in excess of $200 thousand per
     kilowatt.

2.   Solar evaporation is used to obtain salts from seawater or brines —
     This traditional method continues to be important on both large
     and small scales in many parts of the world, including the U.S.
     where a million tons of salt is produced annually by evaporation.

3.   Solar distillation of water is used on a limited commercial scale —
     Desalinization processes are in pilot-plant stages of development.
     Small communities in some isolated areas depend on solar
     distillation for fresh water supplies.

4.   Solar water heaters are used in several countries — Fairly signifi-
     cant numbers of water heaters for individual dwellings are used in
     the United States, Israel, Australia, and Japan.

5.   Solar energy is used for drying materials — This traditional
     application (especially to agricultural products) is of substantial
     importance in many areas. Some people even continue to dry
     laundry this way.

6.   Significant individualized applications have been made — Homes
     using solar heating and cooling (usually to augment conventional
     systems) have been built, but not on a widespread, commercial
     basis. A high-temperature solar furnace has been in operation in
     France for a number of years.

Collection and Storage Systems

If solar energy is to become commercially available and widely
used, technological advances must be made in systems used for its
collection (conversion) and storage. Most proposed or currently used
collection systems convert sunlight into either heat or electricity.
Collectors, now used to convert sunlight directly into heat, are typically
made up of a black, sunlight-absorbing metal surface that is covered
with several layers of glass to minimize radiative heat loss. The effectiveness of such systems is enhanced by the greenhouse effect. The glass
is transparent to the sunlight of shorter wavelengths but absorbs and
traps the longer wavelengths radiated back by the hot surface. A fluid
circulated through pipes in contact with the hot surface is heated and
used to carry the heat to its point of use. This system, illustrated in
Figure 9—16A, is best suited to applications such as water heating and
comfort heating or cooling of residences.

In another approach, it is suggested that solar energy be used to
generate high-temperature steam which could drive a turbine electrical
generator. The high steam temperatures needed for efficient turbine
operation cannot be attained with the simple system described above.

They can be produced only by concentrating the sunlight and reducing the energy losses from the collector. These conditions can be met by using lenses or reflectors to concentrate the sunlight and specially designed thin films to greatly enhance the energy-absorbing and -retaining properties of the collector. Thin films with the necessary selective radiation-transmitting properties are available, but their useful lifetimes at the proposed high operating temperatures are uncertain and must be further evaluated. This collector system is shown in Figure 9—16B.

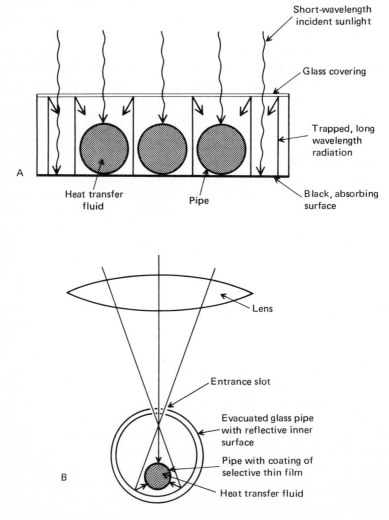

Figure 9—16.   Collector Systems for Converting Sunlight into Heat

Modified with permission from "Physics Looks at Solar Energy" by Aden Baker and Marjorie Pettit Meinel in *Physics Today* (Feb. 1972): Fig. 6, p. 49. Reprinted with permission from the American Institute of Physics, Inc.

The direct conversion of solar energy into electricity can be accomplished with photovoltaic or so-called solar cells. These useful devices were invented at the Bell Telephone Laboratories in 1954 and consist of two thin layers of specially treated silicon sandwiched together. When light strikes the cell, electrons flow from one layer to the other and create a small electrical potential or voltage. Light meters and the self-adjusting cameras used in photography are familiar examples of solar cell use. Silicon solar cells with conversion efficiencies of about 10% are available. Unfortunately, these fragile devices are produced by hand-crafting techniques which make them very expensive and therefore useful only where small amounts of electricity are needed or cost is a secondary concern, as in a spacecraft. Other types of solar cells are being developed which promise to be much less costly and more efficient. Some people believe that silicon cells will be mass-produced at lower costs in the reasonably near future as well. If solar cell technology develops in these directions, the direct conversion of sunlight into electricity could become economically competitive.

The intermittent nature of solar energy makes efficient storage systems a necessity. Various approaches to the problem have been proposed and some have been used. Storage systems range from the simple to the very elaborate. Thermal energy is usually stored in a heat reservoir or heat sink as simple as an insulated pit containing rocks or as complex as a highly insulated tank of liquid metal or salts. The basic operating principle is the same regardless of the nature of the heat sink. Heat is absorbed into the sink from fluid heated in the collector during favorable energy-collecting hours. When the stored heat is needed, another fluid (or in some cases the same one) is heated by circulation through the hot sink. This fluid then carries the heat to the desired point of use. In some heat sinks the heat is stored by raising the temperature of the storage medium. In others, a phase change is used. For example, an appropriate salt mixture or organic solid will melt and absorb the heat of fusion when heated by hot fluid from the collector. A cool fluid circulating through the sink causes the liquid salt or organic compound to solidify, and the heat of fusion is released and heats the cool fluid. A storage-distribution system involving a heat sink is represented in Figure 9—17. Note that the system can also be used for cooling in the summer by removing heat from the interior during the day and storing it in the heat sink until it can be released to the cool outside at night.

Solar energy converted into electricity can be stored in numerous ways, the most obvious being storage cells. In spite of their use in some experimental systems, available storage cells are not considered to be sufficiently durable, maintenance-free, and efficient for commercial systems. Technological improvements in storage cells could solve this problem. Other proposals for electrical storage make use of the pumped-storage concept discussed earlier for hydroelectric power

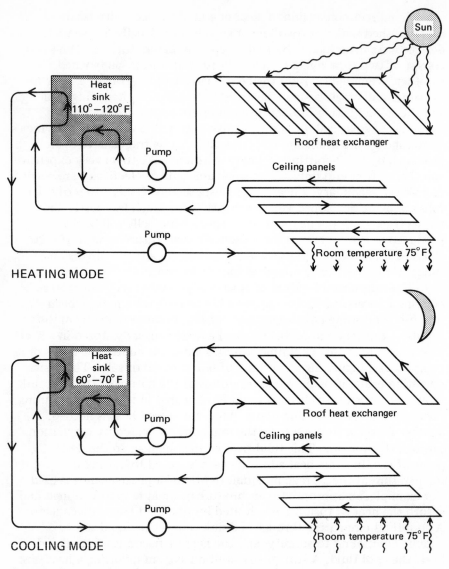

Figure 9–17.   Storage of Thermal Energy from the Sun

Modified with permission from *Introduction to the Utilization of Solar Energy* by A. M. Zarem and Duane D. Erway, eds. (New York: McGraw-Hill Book Company, Inc., 1963), p. 277.

plants. Solar electricity would be used to pump water into reservoirs from which it would be released through hydroelectric generators as needs for electricity developed. Another alternative suggests converting solar electricity into a chemical form from which it would be released as necessary. The electrolytic generation of hydrogen gas from water is an often-mentioned example. The generated hydrogen would be stored (or transported) and used as needed for a fuel. The heat released by burning hydrogen would be used directly or changed back into electricity by steam turbine generating systems or fuel cells. One drawback to such systems is the unavoidable energy losses that accompany each conversion from one form to another. The use of hydrogen gas as a fuel and the topic of fuel cells are discussed in more detail later in this chapter.

Potential for Solar Energy Use

The feasibility of solar energy use is dependent upon a number of factors, including weather patterns, average hours of sunshine daily, and average temperatures. As a result, solar energy could be used only in certain locations of the United States. This is shown in Figure 9—18, where the feasibility of solar heating systems is indicated for areas within the United States. Even within areas where feasibility is projected, seasonal variations in such factors as weather and cloud cover exert significant influences on the usefulness of solar energy collection devices. This is illustrated by Figure 9—19, which shows the monthly variation in the amount of solar energy reaching the ground at three

Figure 9—18.   **Feasibility of Solar Heating Systems in the Conterminous United States**

Redrawn from "Climatic Considerations of Solar Energy for Space Heating: Abstract of a Paper Presented at the Solar Energy Symposium" by T. A. Siple, M.I.T., Aug. 21, 1950 as published in *Energy in the Future* by P. C. Putnam, p. 180. Copyright assigned by D. Van Nostrand Company, Inc., to the General Manager of the Atomic Energy Commission.

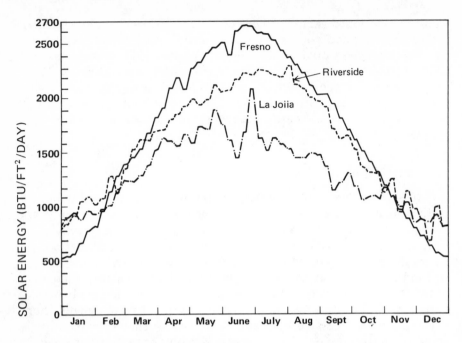

Figure 9–19.   Monthly Variation of Incident Solar Energy, California

Redrawn with permission from *Introduction to the Utilization of Solar Energy* by A. M. Zarem and Duane D. Erway, eds. (New York: McGraw-Hill Book Company, Inc., 1963), p. 35.

locations in California. The lower values for Riverside and La Jolla are caused by increasing cloudiness nearer the ocean. The very high maximum for Fresno was the result of a completely cloudless July.

Solar Energy Development

It is generally agreed that the future development of solar energy on a large scale is going to be so costly that governmental funds will have to be used, at least in part, to finance the necessary research. In a report released in 1973 by a panel of the National Science Foundation and the National Aeronautics and Space Administration, it was recommended that more than $1 billion be spent to develop and construct a solar thermal energy demonstration facility. The total expenditure proposed by the panel for solar energy research was a huge $3.5 billion to be spent over a 10-year period.

The use of solar energy for comfort heating will probably develop in the direction of the small individualized units for each dwelling which we have already discussed. However, the generation of electricity from solar energy is a more difficult challenge, and there are conflicting opinions about the best approach to use. Some believe that small generating units located near the consumer represent the best approach.

Others have proposed large facilities patterned after existing centralized power plants.

The group proposing the small individualized approach favors the use of turbine generators with operating temperatures considerably lower than those used in nuclear or fossil-fueled power plants. These turbines operate typically at temperatures below 200°C. As a result, they have low thermal efficiencies between 10% and 15%. Their economic advantages relative to other sources of electricity have not yet been demonstrated.

Preliminary efforts aimed at the development of large centralized solar power plants are in progress, but significant problems remain to be solved. Several approaches have been suggested, including the use of high-temperature steam turbines driven by heat collected on large areas of collectors based on the design shown in Figure 9—16B. Typically, the collectors for a 1000-megawatt installation will occupy about 12 square miles. This approach will probably require the construction of lens or mirror systems larger than any yet built, and the development of efficient methods for transferring heat from the widespread collectors to the generating facility.

The feasibility of another proposal depends upon the development and production of efficient, low-cost solar cells. Such cells would be arranged in large arrays and their combined output fed to a central point from which it would be distributed to consumers. If a 10% conversion efficiency is assumed, about 14 square miles of collectors would be required for a 1000-megawatt installation.

In a related approach, a satellite containing a large array of solar cells would be put into a synchronous orbit high above the earth. The energy from the collector would be transmitted to earth as microwave radiation which would then be converted back into electricity. This type of system, represented in Figure 9—20, could probably not be attempted before the turn of the century. Its success depends on space technology as well as solar cell improvements.

Environmental Problems with Solar Energy Use

Some problems related to solar energy development have already been discussed. They include the necessity for technological advances in a number of areas and high developmental costs. In addition, some potential environmentally related problems have been identified. The large areas required for the collector systems of large centralized installations is disturbing to some. However, proponents point out that the areas involved are about the same as those now leased to strip mine coal which will produce an equivalent amount of energy. Furthermore, the solar energy farms might also be put to secondary uses such as cattle grazing. Thermal pollution is another concern for installations

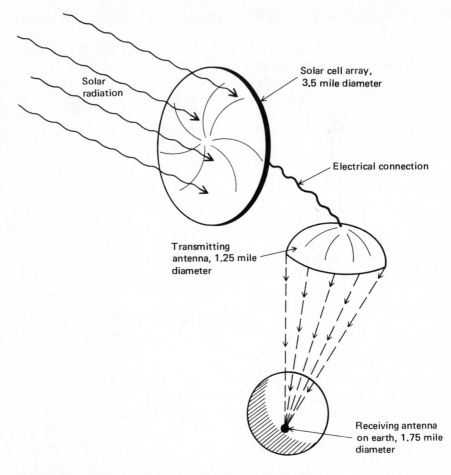

Figure 9–20.   Conceptual Design for a Satellite Solar Energy System

Adapted from "AEC Authorizing Legislation, Fiscal Year 1973," Subcommittee Hearings U.S. Congress, Part 2, Feb. 22–23, 1972, p. 1098.

using steam turbine generators. This problem exists with any thermal conversion system but can be minimized by proper precautions. Concern has also been expressed about the possibility of upsetting the climate in the vicinity of huge solar energy collecting systems. The net energy balance of the earth is changed by absorbing large amounts of solar energy that would otherwise be reflected or radiated back into space. The net balance might be restored by "painting" areas near the collectors white to reflect and emit radiation, thus offsetting the "blackness" of the collector.

Energy from the Wind

Windmills have been in use for many years as small-scale sources of mechanical energy — more than 30,000 were operating in Europe during the nineteenth century. The first wind-generated electricity was produced in 1890. In the present century, however, wind-powered equipment with its intermittent output has largely been replaced by devices dependent on fossil fuels. Current energy shortages and progress in the design of more efficient wind-powered equipment have rekindled an interest in wind-driven devices, especially electricity-generating turbines (aerogenerators).

The use of wind power has several obvious advantages. The wind is free, the delivery of wind to the site of use is free, the supply of wind is inexhaustible, and the use of wind releases no detrimental pollutants into the environment. The disadvantages of wind use are that the energy of wind is diffuse, wind velocities are variable and unpredictable, wind energy is delivered intermittently, aerogenerators have low conversion efficiencies for changing wind energy into electricity, and icing as well as other severe weather conditions adversely affect aerogenerators.

Requirements for Development

Numerous geographical areas appear to be suitable for aerogenerator use, but such use will depend upon a number of favorable circumstances which will make wind-generated electricity economical. These circumstances include:

1. A favorable overall wind velocity — The economical production of electricity requires an average annual wind velocity of nearly 30 mph. This velocity and the wind direction must be reasonably steady.
2. Low construction costs — Sites must be amenable to minimal installation and development costs.
3. Use of an electrical distribution system capable of profitably using interruptible power or development of methods to smooth the intermittent output — Output from aerogenerators might be used as a supplemental supply to an electrical system with a peak generating capability large enough to satisfy demands without contributions from the aerogenerators. If the aerogenerator output decreased, the slack could be taken up by the rest of the system. The output of aerogenerators might be used to pump water into storage reservoirs. The water could then be released as needed to flow through hydroelectric generators and produce an uninterrupted supply of electricity.

A new concept in the use of windpower suggests that aerogenerators be mounted at sea on either floating platforms or on platforms resting on the continental shelf. The produced electricity would be used to electrolyze distilled seawater into hydrogen and oxygen. The hydrogen gas would be collected and transported to shore through a pipeline. On shore energy would be released by burning the hydrogen or using it to power fuel cells. The hydrogen could also be compressed and stored for use during periods of peak energy demand.

Wind Power in the United States

No large-scale aerogenerator systems are now operating in the United States. However, from 1941 until 1945 some work was done toward developing such a system. A large structure was built near Rutland, Vermont. It stood 110 feet tall and was topped by stainless steel blades 175 feet in diameter. The blades turned the main shaft at nearly 29 revolutions per minute. Through a gear arrangement, the rotor of the generator was turned at 600 revolutions per minute. This system only operated intermittently but its output of 1250 kilowatts proved that wind power was capable of producing megawatt quantities of electrical power. The project was terminated in 1945 because of structural failure in the blade spars. The inadequacy of the spars was known before failure occurred, but war priorities for materials prevented any changes from being made.

High costs for equipment, energy storage, and backup systems coupled with the intermittent nature of wind-generated electricity seem to preclude any favorable cost benefit for the present development of wind power facilities. A considerable amount of additional research and development is required to bring the costs down. The Office of Science and Technology estimates that about 1% of the nation's electricity demand could be supplied by wind power in the year 2000. This estimate assumes a high level of government support will be made available for technology development and achievement of economic feasibility. Although wind power will never be a major contributor to the total U.S. energy supply, it could become a significant factor in certain regions of the country.

Environmental Impact of Wind Power

The use of wind power could reduce the amount of air and water pollution produced annually (wind generates no such pollutants) and decrease the dependence on energy from polluting nuclear or fossil-fueled plants. The primary environmental disadvantage could be the aesthetic impact of large numbers of towers located in prominent positions. These towers, some possibly 1000 feet high, would probably require guy wire systems as well as elaborate electricity distribution systems. The total effect might prove to be an aesthetic disaster.

Energy from Ocean Thermal Gradients

Jacques d'Arsonval, a French physicist, predicted in 1881 that the oceans of the world would someday be mined for energy to run our civilization. This prediction has not yet become a reality, but three groups of United States researchers are now studying the concepts d'Arsonval proposed. These concepts involve the utilization of temperature differences that exist between seawater at the surface and at lower levels. Such temperature differences are commonly called *ocean thermal gradients*, or OTG.

The technical feasibility of energy extraction from ocean thermal gradients was first demonstrated by Georges Claude of France in 1929. However, the major problem of economical feasibility remains to be solved. Current cost estimates offer hope that OTG energy might be competitive with other types, but such estimates must still be substantiated by actual testing now under serious consideration.

Description of an OTG Plant

Current plans call for OTG plants to be constructed as floating structures with all or most of the mass beneath the ocean surface. The necessary operating parts of the plant include a heat exchanger, a turbine generator, a condenser, and a liquid pressurizer. During operation, warm surface seawater passes through the heat exchanger where it heats and vaporizes a low-boiling working fluid. The vapor from the working fluid expands through a turbine to produce electricity. The working-fluid vapor is then condensed back to a liquid in a condenser cooled by cold seawater brought from below the surface.

A very important component of the process is the working fluid. It must be a substance with good heat transfer characteristics and a reasonably high vapor pressure at the temperature of the surface water. The heat transfer characteristic is needed to minimize the size and cost of heat exchangers and condensers. The high vapor pressure is required to avoid the necessity of unreasonably large turbines. In submerged units the vapor pressure must be higher than atmospheric pressure to compensate for added hydrostatic pressure. The 1927 attempt by Georges Claude to extract OTG energy was an economic failure because he attempted to use the vapor of seawater itself to drive his turbine. Working fluids now being considered include ammonia, propane, and freon.

Figure 9—21 contains a schematic representation of an OTG plant operating between 5° and 25°C (a 20°C gradient) with ammonia as the working fluid.

Location of OTG Plants

The desirable temperature characteristic for OTG energy production is a large temperature difference over a short distance. This characteristic is most often found in tropical ocean water where a temperature difference of 27.5°C (50°F) commonly exists between surface water and water found 2000—5000 feet below the surface.

No tropical oceans are found adjacent to the United States, but the Gulf Stream brings very warm near-tropical surface water near the coast of Florida. The temperature difference between these surface waters and the underlying cold water is always at least 15°C (27°F), and during half the year it reaches 23°—30°C (42°—45°F).

The preliminary designing is underway for a submerged OTG plant to be located in the Gulf Stream about 15 miles from Miami. The current design concept is modular with six turbines in each of two hulls. Each hull would be 480 feet long and 100 feet in diameter. The output of the plant is rated at approximately 400 megawatts.

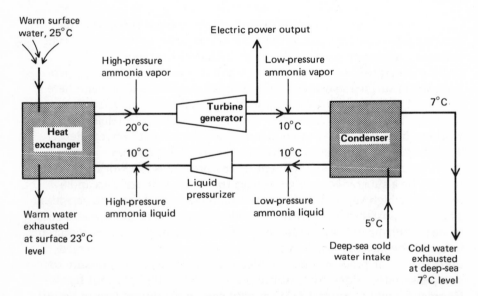

Figure 9—21.   Ocean Thermal Gradient Power Plant Operation

Redrawn with permission from "Solar Sea Power" by Clarence Zener in *Physics Today* (Jan. 1973): Fig. 1, p. 50. Reprinted with permission from American Institute of Physics, Inc.

Efficiency of OTG Plants

The maximum thermal efficiency of an OTG plant is determined by the temperature difference involved in the process (remember equation 2—7). A maximum efficiency of a little more than 5% is possible when a temperature difference of 15°C (27°F) is used and the surface water is at 16.6°C (62°F). The efficiency increases to about 8% when surface temperature reaches 26.6°C (80°F) and the differential is 25°C (45°F). These efficiencies are quite low when compared to conventional fossil-fueled plants that operate with approximately 40% efficiency.

Savings in plant construction partially compensate for the cost handicaps resulting from the low efficiencies of OTG plants. Conventional plants operate at high temperatures and pressures. The heat exchangers must be constructed with thick strong walls of expensive heat-resistant material. An OTG plant can have thin walls in the heat exchanger because of the low pressures involved and the compensating effect of outside hydrostatic pressure (in submerged plants).

The amount of warm seawater that must pass through an OTG plant is tremendous. However, it is no greater than that which passes through a modern hydroelectric plant producing an equivalent amount of energy. The energy extracted from 1 pound of water in an OTG plant utilizing a temperature differential of 20°C is the same as that obtained from 1 pound of water in a hydroelectric plant with an operating head of 93 feet.

Problems Involving
OTG Plants

Both the heat exchanger and condenser in an OTG plant must be exposed to corrosive seawater. The metals used in these parts must therefore resist corrosion. Tests have shown the pure aluminum is suitable because an adherent oxide coating forms on the surface and prevents any further corrosion. High-strength aluminum alloys to which an oxide coating has been pressure bonded would also be suitable for use and would have long service lives in OTG plants.

The locations of OTG plants are generally limited to temperate zones which are usually far removed from energy-hungry population centers. This creates a problem of energy transmission to consumers. Two transmission methods now receiving attention could be used for OTG energy. The first makes use of underground cables, possibly low-temperature superconducting types. The use of superconducting cables for power transmission has recently been studied in great detail. The general conclusion is that such use is feasible if the power to be transmitted is sufficiently high — greater than 40,000 megawatts.

According to a second approach, electricity from the OTG plant would be used to electrolyze water into hydrogen and oxygen gas. The gaseous hydrogen would then be transferred to shore through a pipeline or possibly liquefied and transported to shore in cryogenic tankers similar to those used to ship liquefied natural gas. On shore the hydrogen could be burned as a fuel or stored for future use. This was discussed previously in conjunction with offshore wind power. The use of hydrogen as a fuel is discussed in detail later in this chapter.

Environmental Effects of OTG Plants

The effects of OTG plant operations on the environment have been the subject of little research. The initial analyses indicate that the primary effect would be an increase in the amount of thermal energy stored in the ocean. This results because of the discharge of larger-than-normal amounts of colder water near the surface. This colder water would absorb energy from the air and sun until it reached the temperature of the surrounding water.

The possible depletion of cold water is no problem, since it is continually replenished through a natural cycle. The surface waters of the oceans are heated in the tropics. This warmed water flows toward the poles and gradually gives up its heat. Some ice in the polar regions is melted and the warm surface water is cooled. The cooler water is more dense and gradually sinks to the bottom where natural currents slowly carry it back to the equator.

Secondary Chemical Fuels

It is now being acknowledged that nonrenewable fossil fuels, the main sources of energy today, will inevitably be used up. As a result, a great deal of research and development is being done relative to alternate energy sources. The alternates are likely to be nuclear (fission and fusion), solar, geothermal, tidal, and wind energy. The previous discussions about these sources show one bit of commonality — in most cases the energy appears as electricity. This implies that electricity is the form in which energy from these sources will be used. However, an alternative exists: electricity can be used to generate secondary chemical fuels which can be used in place of primary fuels. Research into this concept has generally focused attention on hydrogen gas as a likely secondary fuel, and it will therefore be used as an illustrative example. Much of the discussion is quite speculative, and a great deal more research and development will be needed to bring a hydrogen fuel economy into existence.

The necessity for developing a secondary chemical fuel becomes clear when the potential problems associated with an all-electric or predominantly electric economy are reviewed. The following major problems have been identified:

1. Plant power location — The requirement for power plant cooling facilities (either cooling water or atmospheric cooling towers) for both fossil-fueled and nuclear power plants plus safety concerns about nuclear plants will limit the number of suitable sites available for power plant construction. The resulting remote locations will create problems associated with energy transmission methods and costs, and energy storage.
2. Energy transmission — The expense involved in transmitting large amounts of electrical energy increases rapidly as the amounts of energy get larger. The necessary high-voltage overhead lines become increasingly costly in terms of both the equipment and the land (rights-of-way) needed. In addition, they are vulnerable to storm damage and considered to be esthetically undesirable. Electrical utilities are encountering increasing resistance to the continued use of overhead transmission lines in many areas of the country. High-capacity underground cables cost at least 9 and sometimes 20 times as much as overhead lines and thus are far too expensive for long distance power transmission.
3. Energy storage — Electricity cannot be stored conveniently; it must be produced at almost exactly the consumption rate. Pumped water storage, discussed earlier, is a practical alternative but is limited by the availability of suitable sites. Therefore, most electrical supply systems must be expensively over-designed to accommodate fluctuations in demand.
4. Heat energy use — Most energy is used in the form of heat (from combustion) rather than electricity. Some users of combustion heat could convert to electricity, but for many electricity is simply not suitable. For example, can you visualize an electric airliner?

A secondary chemical fuel could ease these problems of the all-electric economy. It would be combustible and could therefore be used for supplying heat energy; it could be stored — permitting generation under the most favorable conditions rather than as dictated by demand; and it could be transported easily over long distances at what are estimated to be lower costs compared to electricity.

Hydrogen as the Secondary Fuel

Hydrogen is considered to be a suitable and desirable secondary chemical fuel for a number of reasons, including the following:

1. Hydrogen is essentially a nonpolluting fuel — No carbon monoxide, sulfur dioxide, hydrocarbons, or particulates are

generated during the combustion of hydrogen. The only significant product is water, as the following reaction indicates:

$$2H_2 + O_2 \longrightarrow 2H_2O \qquad\qquad 9-1$$

Small amounts of nitrogen oxides are formed when air is used in any combustion process. However, the amounts generated during hydrogen combustion are generally lower than the amounts produced by other fuels.

2. The source of the hydrogen is essentially unlimited — The raw material for hydrogen production is anticipated to be water. More information about the reactions involved is given later in this chapter.

3. The replenishment of the hydrogen source is rapid — Water, used to generate hydrogen, is produced during combustion. Thus, the cycle involving raw material and fuel is short and direct. In areas with small water supplies, the cycle could be self-contained and the water used over and over. The short renewal time for the hydrogen raw material (compared to fossil fuels) is shown in Figure 9—22. Note that the generation of secondary fuel requires energy produced from a primary or naturally occurring fuel.

Other potential secondary fuels have some of the advantages of hydrogen but none have them all — particularly the first one. Other fuels that have been considered are alcohols, hydrazine, and ammonia. All of these can be produced from the readily available raw materials water, nitrogen, and carbon dioxide.

The idea of using hydrogen as a secondary fuel is not new. In 1933 Rudolf A. Erren, a German inventor working in England, suggested that off-peak excess electricity be used to produce hydrogen on a large scale. He had worked extensively at modifying internal combustion engines to burn hydrogen. Interestingly, his main goals were to eliminate air pollution from auto exhaust and decrease Britain's dependence on petroleum imports.

Current Production and Use of Hydrogen

Hydrogen is a widely used substance in many industrial processes. Its properties have been studied extensively and are well known. Consequently, its use as a secondary fuel would not require the development of new techniques for handling, transporting, or working with it.

Currently, hydrogen consumption in the United States amounts to more than 2.3 trillion cubic feet per year. Annual world consumption stands at about 6 trillion cubic feet. Most of the current hydrogen

PRESENT ENERGY PRODUCTION FAR OUTPACES TIME FOR RENEWAL . . .

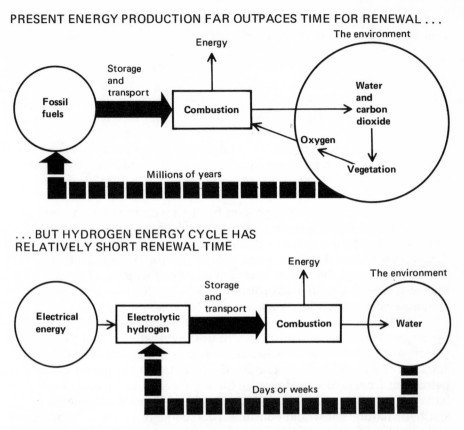

Figure 9—22.   Renewal Times for Hydrogen and Fossil Fuels

Redrawn with permission from *Chemical and Engineering News,* 50 (June 26, 1972): 14—15.
Copyright by the American Chemical Society.

production is used as a chemical raw material rather than as a fuel. Most
of it is used in two specific industrial applications: ammonia production
(42%) and petroleum refining (38%). Ammonia, the fundamental
compound for the fertilizer industry, is made by reacting hydrogen
with nitrogen as follows:

$$N_2 + 3H_2 \xrightarrow[\text{pressure}]{\text{heat, catalyst}} 2NH_3 \qquad\qquad 9\text{--}2$$

Large amounts of hydrogen are used in petroleum refining to improve
the quality of refined products and to desulfurize crude petroleum. The
remaining 20% is divided among many industries, with metallurgical
and food processing applications taking the largest share.

The primary industrial processes used to prepare hydrogen are variations of the catalytic steam hydrocarbon reformation process represented by the following reactions:

$$CH_4 + H_2O \xrightarrow[\text{heat}]{\text{catalyst}} CO + 3H_2 \qquad\qquad \textbf{9-3}$$

$$CO + H_2O \longrightarrow CO_2 + H_2 \qquad\qquad \textbf{9-4}$$

$$CH_4 + 2H_2O \longrightarrow CO_2 + 4H_2 \qquad\qquad \textbf{9-5}$$

In the process, hydrocarbons are catalytically reacted with steam at moderate temperatures. The products are carbon oxides and hydrogen. The carbon dioxide is removed from the hydrogen in another step of the process.

The only currently used process for hydrogen production that does not necessarily consume hydrocarbons is the electrolysis of water. The passage of a direct electrical current through water causes it to decompose directly into hydrogen and oxygen (see Figure 2—3). A low-cost source of electricity is needed to make this process competitive with those based on hydrocarbon reforming. Such sources could become available if fusion reactors prove feasible.

The demand for hydrogen will increase in the future whether hydrogen becomes a secondary fuel or not. Synthetic gas and oil industries of the future (see Chapters 6 and 7) will consume large amounts of hydrogen. For example, the production of synthetic crude petroleum from coal would require 6500 cubic feet of hydrogen per barrel of petroleum manufactured, and coal gasification to generate synthetic natural gas would consume 1500 cubic feet of hydrogen for each 1000 cubic feet of SNG formed.

Future Hydrogen Production

If hydrogen is to replace fossil fuels on a large scale in the future, tremendous amounts will be required. Approximately 300 trillion cubic feet per year would be required by the year 2000 to replace all fossil-fuel use for everything except electricity generation. About 60 trillion cubic feet would be required just to replace current annual U.S. natural gas consumption. The significance of these amounts becomes apparent when we realize that present U.S. production is less than 3 trillion cubic feet per year.

New or improved generation methods will be required if such demands are to be met. The consensus of those working on the problem is that water is the only feasible source for the long term. Most ideas for large-scale production involve a nuclear power source and water; at least three techniques are being considered: (1) electrolysis of water, (2) thermal decomposition of water, and (3) photolysis of water. Of these three techniques, only electrolysis has been proven feasible in practice. Neither of the other two is past the pilot plant stage of development.

Decreased costs hold the key to the future acceptability of electrolytically generated hydrogen. Low cost electricity and higher conversion efficiencies are necessary to achieve the desired goal. The efficiencies of commercial electrolyzers now in operation are between 60% and 70%. Among other things, it is hoped that applications of aerospace technology will improve this figure. One promising break-through involves the use of solid polymer electrolytes originally developed for use in the fuel cells of space vehicles. The SPE is a specially formulated plastic sheet that has many of the properties of teflon. None of the three acids or alkalies characteristic of conventional commercial electrolysis cells are required. Ionic conductivity is provided by the movement of hydrated hydrogen ions through the sheet of SPE.

The direct thermal decomposition of water avoids the inefficient process of converting heat into electricity. Nuclear reactors have been proposed as the heat sources. The simple-appearing reaction is:

$$2H_2O \xrightarrow{\text{heat}} 2H_2 + O_2 \qquad\qquad 9-6$$

However, no appreciable reaction takes place at temperatures below $2500°C$, and commercial nuclear reactors are not expected to be sources of useful temperatures much higher than $900°C$. For this reason, the one-step thermal decomposition process has essentially been discarded.

A multistep process carried out at lower temperatures might be possible. Research concerning such processes is being conducted in a number of laboratories. The Euratom research center in Italy has developed a promising four-stage sequence in which the dissociation of water is carried out at temperatures below $730°C$. The steps are:

$$CaBr_2 + 2H_2O \longrightarrow Ca(OH)_2 + 2HBr \qquad\qquad 9-7$$

$$Hg + 2HBr \longrightarrow HgBr_2 + H_2 \qquad\qquad 9-8$$

$$HgBr_2 + Ca(OH)_2 \longrightarrow CaBr_2 + HgO + H_2O \qquad\qquad 9-9$$

$$HgO \longrightarrow Hg + \tfrac{1}{2}O_2 \qquad\qquad 9-10$$

The net reaction for the process is simply the dissociation of water, since none of the other materials involved are consumed. Problems that remain to be solved are related to the effects of highly corrosive compounds such as hot hydrobromic acid (HBr) and the poisonous nature of the mercury compounds used. The process appears to be feasible, but more developmental work is needed to bring costs down and increase efficiencies to practical limits.

For the more distant future a third type of process may be possible. The process, proposed by Bernard L. Eastlund of the Atomic Energy Commission, would use ultraviolet radiation from the plasma of a fusion reactor to photolytically decompose water. He says the injection of heavy metals such as aluminum into a hydrogen plasma would produce radiation of the correct wavelength for photolysis. The photolysis would take place in water vapor circulating around the plasma, as represented in Figure 9—23. The experimental confirmation of Eastlund's scheme will require the development of an operating fusion reactor, but it seems possible that such a process, in conjunction with electrical generation, could increase the efficiency of a fusion power plant.

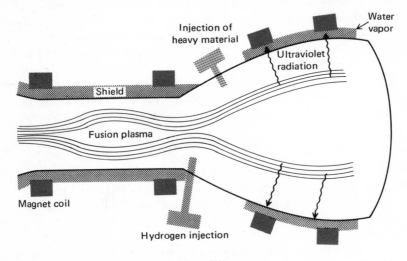

Figure 9—23.   The Photolytic Decomposition of Water

Redrawn with permission from *Chemical and Engineering News,* 50 (July 3, 1972): 17. Copyright by the American Chemical Society.

Transmission and Storage of Hydrogen

The wide use of hydrogen requires the availability of efficient transmission systems. Since hydrogen is a gas, it can be transmitted and distributed by pipelines just as natural gas is handled today. Pipelines have proved to be one of the least expensive methods for moving fuels (both liquids and gases). In some refinery operations, hydrogen is now routinely transported in pipelines, although only over relatively short distances.

Hydrogen carries only about one third as much energy per cubic foot as natural gas, but it is much less dense and flows about 3 times as easily. Therefore, a pipe of fixed dimension would have nearly equal energy-carrying capacities for either substance. Increases in the pumping capacity of pipeline systems would be required in order to move the larger volume of hydrogen equivalent to the natural gas. This would increase transmission costs to a level of about 120% those of natural gas. This increase is not considered to be prohibitive since the transmission costs for hydrogen are still proportionately smaller in relation to total costs than they are for natural gas, as shown by Figure 9—24. Note that the production cost of hydrogen produced from electricity is higher than that of electricity — this will always be the case. It is the lower transmission and distribution costs of hydrogen compared to electricity that make it advantageous to the consumer. As an alternative to pipeline transmission, hydrogen can also be transported in liquid form. Trucks and ships are already performing this task, and the

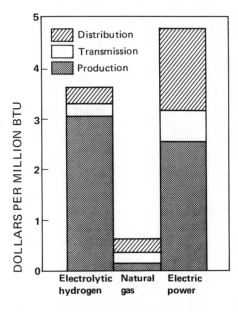

Figure 9—24.   Relationship of Transmission Costs to Total Costs

Redrawn with permission from "Hydrogen Fuel Economy: Wide-Ranging Changes" in *Chemical and Engineering News*, 50 (July 10, 1972): 27. Copyright by the American Chemical Society.

technology of handling cryogenic materials is well established, particularly in the aerospace industry. Another alternative to pipelines is the conversion of hydrogen into chemical forms, such as ammonia (NH_3), methanol (CH_3OH), or cyclohexane (C_6H_{12}), from which hydrogen can be regenerated at the consuming facility.

The storage capability of hydrogen is extremely important in order to even out daily and seasonal demand variations. Natural gas is stored in huge quantities in underground porous rock formations and in depleted gas fields. The applicability of this technique to hydrogen storage can be determined only by future field trials. However, at the present time, large volumes of helium, a low-density gas with leak characteristics similar to hydrogen, are satisfactorily stored in an underground reservoir located near Amarillo, Texas. In areas where underground storage is geologically impossible, a recent trend has been to liquefy natural gas and store it in refrigerated tanks. The aerospace program has generated the technology needed for this process. A liquid hydrogen storage tank with nearly a million-gallon capacity is in use at the Kennedy Space Center in Florida. The storage costs for liquid hydrogen are 2 to 3 times greater than for liquid natural gas, but they might be reduced by large-scale engineering techniques. It is worth noting that the storage costs for pumped hydroelectric energy are many times higher than those of liquid hydrogen.

Metal hydrides have been considered as another possible storage mode. The characteristics of hydrogen allow it to easily penetrate into and combine with various metals and alloys. In some metals, such as titanium, the process is so complete that the resulting hydrogen concentration per unit volume is greater than in liquid hydrogen. The hydrides are formed by exposing the metals to pressurized hydrogen. Heat is given up in the process, which can be reversed (and the hydrogen released) by heating. A pilot unit based on an iron-titanium hydride is part of a test facility being planned by the Public Service and Gas Company and Brookhaven National Laboratories. This facility will be used to satisfy peak energy demand requirements.

Hydrogen Utilization

The use of hydrogen as a fuel presents some problems together with some distinct technical advantages. There seems to be no reason why pure hydrogen could not replace natural gas in all present uses. When mixed with air, it burns smoothly and easily on burners very similar to natural gas burners. Necessary modifications to natural gas burners can be made easily. The only undesirable products of hydrogen and air combustion are nitrogen oxides. These can be eliminated if combustion temperatures are lowered, a process that may be possible by using catalysts. Simple domestic heating and cooking devices have been demonstrated which are based on the use of a catalyst bed in which flameless fuel oxidation goes on.

Transportation is an obvious area for the potential use of hydrogen fuel because of the need for a portable supply of a nonpolluting fuel. The use of hydrogen as a direct fuel and in fuel cells has already been demonstrated in a number of experimental vehicles. The major impediment to using hydrogen as a vehicle fuel is the need for an adequate fuel storage capacity. Gaseous hydrogen, even when highly pressurized, is far too bulky and requires the use of very heavy containers. Liquid hydrogen storage is too expensive at the present time, but technically feasible. A promising possibility is the use of the previously mentioned hydrides which can be decomposed by the heat of the exhaust to liberate pure hydrogen. The early stages of research into this possibility are underway at the Brookhaven National Laboratory.

Liquid hydrogen has some definite advantages as an aircraft fuel. It is lightweight and has an energy-to-weight ratio about 2.5 times as large as conventional hydrocarbon fuels. In addition, it has an inherent cooling capacity about 30 times that of conventional jet fuel. This characteristic could be used to provide cooling for engine parts or aircraft skin surfaces. Some disadvantages also exist for hydrogen as an aircraft fuel. The primary problem is caused by the requirement of on-board cryogenic storage. This makes it necessary to insulate fuel tanks and develop materials that maintain their dimensional stability and strength even when subjected to large temperature differentials. Aircraft designers think these difficulties can be overcome.

The availability of large amounts of low-priced hydrogen would have a distinct influence on industry. For example, the use of hydrogen as a direct reducing agent for the production of iron from ore has been technically proven. The switch from coke to hydrogen in the steel-making industry would have widespread effects, especially in reducing air pollution.

One of the side effects of a future large-scale hydrogen economy would be the availability of large amounts of oxygen produced as a by-product during the electrolysis of water. A wide variety of new or expanded uses for oxygen would be made possible, or maybe even necessary in order to improve the economics of hydrogen production. Oxygen might be used to support combustion in applications where atmospheric gases are not desirable. Large quantities would definitely be used in the production of steel. An even larger market might develop in the treatment of raw sewage.

In order to put all of the above details concerning hydrogen use into proper perspective, it should again be noted that none of the applications is economically realistic unless new hydrogen-producing technology is developed. Current methods could not supply the vast amount needed.

The Safety of Hydrogen

Safety is perhaps the most controversial issue surrounding the possible use of hydrogen as a fuel. Skeptics often express fear about the flammability of hydrogen. Proponents of hydrogen use point out that this "Hindenburg Syndrome" does not take into account the improvements in hydrogen technology and safety measures that have been made since 1937 when the zeppelin disaster took place. Many of the improvements have been spin-off benefits of the space program. Today, hydrogen is considered to be no more hazardous to use, transport, and store than gasoline or natural gas, provided proper equipment and techniques are employed.

The flammability range of hydrogen and air mixtures is very wide, from 4% to 75% hydrogen. Natural gas-air mixtures by comparison are flammable only when the natural gas is present in concentrations of 5% to 15%. However, the crucial characteristic is the lower limit, since it is associated with fire hazards caused by leaks. The lower limits are practically the same for the two fuels, but the higher diffusion rate of hydrogen and its very low density tend to make it safer than natural gas. Hydrogen will rise through air and dissipate more rapidly than natural gas. The most significant property of hydrogen-air mixtures is the extremely low energy (temperature) needed to ignite the mixture — it is only one-tenth that needed to ignite gasoline vapor-air or natural gas-air mixtures. Safety precautions regarding this property are essential if hydrogen is used for any purpose.

Hydrogen has no odor, a property that makes leaks of pure hydrogen hard to detect. Odorants are normally added to natural gas to make leaks more obvious, and the same thing can be done with hydrogen. The nearly invisible hydrogen flame might be dangerous as well; an illuminant added to the gas to increase the flame luminosity would solve the problem.

The dilution of hydrogen with another gas has been suggested as a method for decreasing any hazards associated with hydrogen. Ideally, the diluent would add to the heating value and would produce few undesirable materials upon combustion. Natural gas has been considered as a possible diluent. An alternative to dilution would be to use concentric distribution pipes containing hydrogen in the center pipe. The hydrogen would be isolated from the atmosphere by a pressurized gas flowing through the outer pipe.

Improved Energy Conversion Efficiencies

While the search for new energy sources goes on, other scientists are working on methods for improving the efficiency of processes used to change energy from one form to another. As we have seen, such changes cannot be carried out with anything approaching 100% efficiency. The change of combustion- or nuclear-generated heat energy into electricity is one of the more common processes carried out. Available evidence indicates that technology now being developed will greatly improve the efficiencies of these conversions. Two promising techniques in this category, which we will discuss, are the use of fuel cells and magnetohydrodynamics (MHD).

Fuel Cells

Fuel cells are electrochemical devices with the capability of converting the chemical energy of fuels such as hydrocarbons or hydrogen directly into low-voltage direct-current electricity. They contain electrodes and an electrolyte just as batteries do, but unlike batteries, fuel cells are open systems which require and utilize a continuous supply of reactants (fuel). As long as fuel is supplied, electricity is generated; they do not become "discharged."

Fuel cells have received considerable attention because of the following advantages they offer:

1. They emit few air pollutants — The fuel does not undergo combustion but is converted directly into electricity.
2. They operate quietly — The only moving parts are those in the fuel supply and cell-cooling systems.
3. Their conversion efficiencies are high at low power output levels — The efficiencies of the best steam and gas turbine generators are highest when power output levels are high (greater than 100 megawatts). Fuel cells may be only slightly more efficient but the efficiency is possible at outputs as low as 25 kilowatts (see Figure 9—25A).
4. Their efficiencies are high at less than maximum output levels — Conventional generators are less efficient when they are operated under partial load (see Figure 9—25B).

The fuel cell concept was first suggested in 1802, and in 1839 the first laboratory unit was demonstrated. Continued attention was given to their development from 1900 to 1930, when the improving internal combustion engine discouraged further work. The space program generated renewed interest, and in 1959 the first practical fuel cell was demonstrated. Since that time, fuel cells have been used extensively in the United States space program, but their prohibitive costs have effectively precluded their use as earth-bound power sources. It has only recently begun to appear that the cost problems can be overcome and fuel cells made commercially practical.

The hydrogen-oxygen fuel cell is the best-known type. It was the first to be developed with practical power levels (5 kilowatts in 1959), and it has been used extensively in both the Gemini and Apollo spacecraft. Several types have been developed, but they all contain the fundamental components shown in simplified form in Figure 9—26.

The negative side of this cell is the hydrogen electrode (the anode) where pure hydrogen gas diffuses through a porous material and reacts with hydroxide ions to produce water and liberate electrons.

$$2H_2 + 4OH^- \longrightarrow 4H_2O + 4e^- \qquad 9\text{--}11$$

A catalyst embedded in the porous electrode causes the reaction to take place at moderate temperatures. The electrons flow through the external circuit to the positive electrode (cathode) where oxygen gas diffuses through the catalyst-containing, porous electrode and reacts as follows:

$$O_2 + 2H_2O + 4e^- \longrightarrow 4OH^- \qquad 9\text{--}12$$

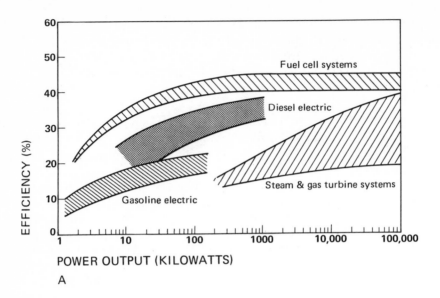

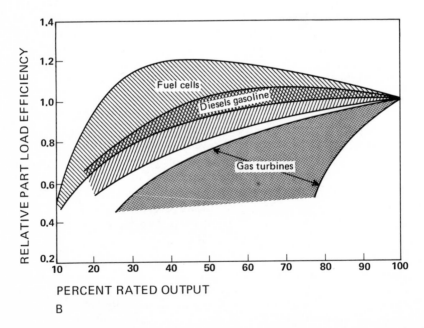

Figure 9—25.   Desirable Characteristics of Fuel Cells   (A) High efficiency at all output levels,
(B) high efficiency at part load.

Redrawn from "Energy Research and Development," House of Representatives Subcommittee
Hearings, no. 24, May 1972, p. 590.

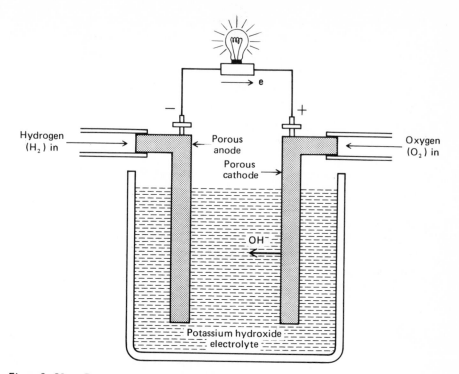

Figure 9—26.   Representation of a Simple Fuel Cell

Redrawn from *Fuel Cells: A Survey* by Bernard J. Crowe, NASA SP—5115 (Washington, D.C.: Government Printing Office, 1973), p. 2.

The generated hydroxide ions complete the electrical circuit by migrating through the electrolyte to the hydrogen electrode.

Fuel cells can be adapted to use a variety of fuels by changing the catalyst, but none have been found to be as efficient as those using hydrogen and oxygen. Other fuel cell systems being studied and developed are:

1.   Reformer-supplied hydrogen and air cells — A series of fuel cells has been developed in which low-cost hydrocarbon fuels are used to generate the hydrogen fuel by a reforming process (see reactions 9—3 through 9—5) carried out in an adjoining apparatus.

2.   Hydrocarbon-air cells — Many hydrocarbons, including the major constituents of diesel fuel, have been successfully oxidized electrochemically in fuel cells, using air as the oxidizer. The reactions have been more than 99% complete, but expensive platinum is the only catalyst found to be suitable. In addition, the power densities achieved to date are only about one tenth that of hydrogen-oxygen cells.

3. Natural gas-air cells — The efficient electrochemical conversion of
    natural gas directly into electricity is difficult because of the
    unreactive characteristics of methane, the main constituent of the
    natural gas. A cell has been developed to use as a fuel the mixture
    of hydrogen and carbon monoxide produced by reacting natural
    gas with steam in the presence of a nickel catalyst.

$$CH_4 + H_2O \xrightarrow[\text{Ni cat}]{500°-750°C} 3H_2 + CO \qquad 9-13$$

The cell, shown in Figure 9—27, operates at moderately high
temperatures and contains a molten electrolyte made up of a
mixture of magnesium oxide, lithium carbonate, sodium carbonate,
and potassium carbonate. A thin layer of porous nickel serves as
the anode, and the cathode is made of finely divided silver. The
reactions that take place in the cell are:
Cathode reaction,

$$2CO_2 + O_2 + 4e^- \longrightarrow 2CO_3{}^{2-} \qquad 9-14$$

Anode reactions,

$$CO + CO_3{}^{2-} \longrightarrow 2CO_2 + 2e^- \qquad 9-15$$

$$H_2 + CO_3{}^{2-} \longrightarrow CO_2 + H_2O + 2e^- \qquad 9-16$$

Fuel cells could be used in a number of significant ways in the
United States energy system. For example, they could be used as
electrical storage systems. During periods of minimum demand the
electrical output of nuclear generators would be converted into
electrolytic hydrogen. The hydrogen would be stored and used in fuel
cells as needed to provide electricity during periods of high demand.
This approach would significantly lower the necessary maximum
capacity of the nuclear generating plant. Studies have shown that the
fuel cell storage concept requires only 1% of the land area needed by an
equivalent pumped-storage hydroelectric system.

Fuel cells may be placed in nearly any location without seriously
disrupting or otherwise affecting the environment. In addition, construction times are relatively short. These characteristics could allow
fuel cells to be used as high-output power plants (25 to 100 megawatts)
to supplement central plant energy production or as low-output plants
(10 to 200 kilowatts) to provide localized electrical power sources for
developing rural areas. The high-output plants would be located at or
near the consuming area, and fuel would be provided by truck or
pipeline. This would eliminate the necessity for transmission lines from
the central plant and reduce the related problems of rights-of-way and
visual pollution.

Fuel cells could also be used by natural gas customers for the
on-site conversion of natural gas into electricity. On the basis of cost

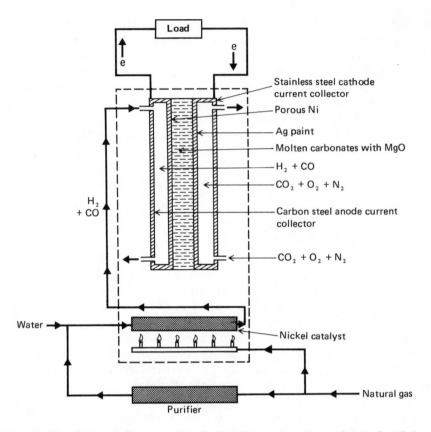

Load

e

e

Stainless steel cathode
current collector

Porous Ni

Ag paint

Molten carbonates with MgO

$H_2 + CO$

$CO_2 + O_2 + N_2$

Carbon steel anode current
collector

$CO_2 + O_2 + N_2$

$H_2 + CO$

Water

Nickel catalyst

Natural gas

Purifier

Figure 9—27.   Schematic Representation of a High-Temperature Natural Gas-Air Fuel Cell

Redrawn with permission from *Modern Electrochemistry*, vol. 2, by John O'M Bockris and
Amulya K. N. Reddy (New York: Plenum Publishing Corporation, 1970), p. 1394.

per unit of energy, transmission and distribution expenses for natural
gas are only 20% to 30% as high as for electricity. In addition to
transmission cost savings, on-site generation of electricity is not as
susceptible to problems resulting from natural disasters or deliberate
damage.

About $200 million were spent between 1967 and 1973 in
attempts to prove the feasibility of commercial fuel cells. A large part
of this money came from a group of natural gas and utility companies
involved in a project known as TARGET — Teams to Advance Research
for Gas Energy Transformation. The goal of the TARGET project is to
produce electricity at a cost competitive with that from large central
generating facilities. As a result of the TARGET project, nearly sixty
12.5-kilowatt natural gas fuel cell power plants have been field tested at
37 locations in the United States and Canada.

A new cooperative project was started in 1973 with the goal of developing a 26-megawatt natural gas fuel cell power plant. This project is being financed by a number of utility companies and the Pratt and Whitney division of the United Aircraft Corporation. An initial commitment of $49 million was made, and the first deliveries of commercial plants is planned for 1978.

The primary problems facing fuel cell developers are initial cost and service life. It is estimated that, with current technology, plants could be built for $350 to $450 per kilowatt and have service lives of about 16,000 hours. In order to be competitive, the cost should be halved and the service life doubled. These objectives must be met by increasing the power output and by developing less expensive and more stable construction materials.

Magnetohydrodynamics

The phenomenon of magnetohydrodynamics (MHD) is the basis of a promising method for efficiently converting thermal energy from fossil fuels or nuclear reactors into electricity. The operating principle of an MHD generator is similar to that of an ordinary turbine-driven generator. In an ordinary generator, a turbine rotates an electrical conductor in a magnetic field. The motion of the conductor in the field causes an electric current to flow in the conductor (Figure 9—28A). In an MHD generator, a stream of high-temperature, electrically conductive gas is passed through a magnetic field. This also induces a flow of electrical current which can be picked up by electrodes (see Figure 9—28B). Thus, the difference between the two generators is the substitution of a conducting gas for the rotating metallic conductor.

The principles for MHD power generation have been known for more than 100 years, but only recently have technological advances produced systems with the promise of commercial usefulness. Extensive laboratory studies of MHD began in the early 1950s at Cornell University. In 1959 MHD power generation was first demonstrated when the

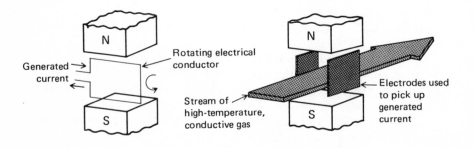

A. Conventional Generator   B. MHD Generator

Figure 9—28.   Comparison of Conventional and MHD Electricity Generators

Mark I generator delivered 10 kilowatts of power. Since then, other units have been developed and demonstrated. The Mark V generator produced 32 megawatts for a short time and was able to provide power for its own electromagnets. The immediate goal of MHD research is the development of a pilot demonstration generator capable of long duration operation at high levels of power output.

Research into MHD power generation is becoming worldwide, with active programs in Japan and several European countries. In the USSR an experimental 75-megawatt MHD power plant is already being tested. The Russian program is primarily oriented toward the use of natural gas as the fuel. Natural gas fuel makes the generator design simpler than that needed when coal, the fuel emphasized in the United States program, is used.

The essential components of an MHD generator are a combustion chamber, a nozzle or channel to direct the flow of hot gases, a powerful magnetic field, and electrodes to collect the electric current. Figure 9—29 is a schematic representation of such a generator. A power plant using an MHD generator would also include a heat exchanger to preheat air fed to the combustion chamber and stack-gas cleaning equipment to remove particulates from the exhaust gases. Preliminary heating of the air used for combustion is necessary to achieve the required operating temperatures unless a large amount of oxygen is added — at present an uneconomical procedure.

MHD generators operate at temperatures of 2000° to 2500°C. These temperatures are not high enough to ionize the combustion gases and make them electrically conductive. The injection of small quantities of seed materials (usually alkali metal salts) overcomes this

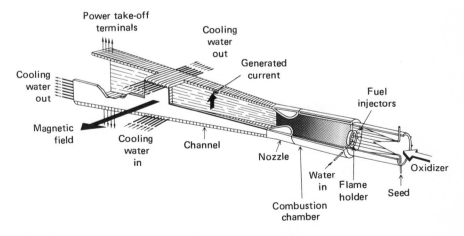

Figure 9—29.   Components of an MHD Generator

Modified with permission from "MHD Power Generation" by Richard J. Rosa from *IEEE Transactions on Plasma Science*, vol. PS—1, no. 1 (March 1973): Fig. 3, p. 8.

problem and produces a conductive gas. The alkali metal compounds
are recovered from the exhaust gases and recycled in the process for
economic reasons. The temperature dependence of the conductivity of
a seeded gas is shown in Figure 9—30.

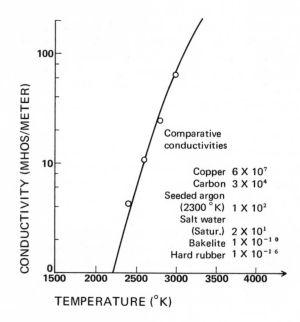

Comparative
conductivities

Copper 6×10^7
Carbon 3×10^4
Seeded argon
(2300 °K) 1×10^2
Salt water
(Satur.) 2×10^1
Bakelite 1×10^{-10}
Hard rubber 1×10^{-16}

CONDUCTIVITY (MHOS/METER)

TEMPERATURE (°K)

Figure 9—30.   Conductivity of Seeded Combustion Products

Redrawn with permission from "MHD Power Generation" by Richard J. Rosa from *IEEE
Transactions on Plasma Science,* vol. PS—1, no. 1 (March 1973): Fig. 2, p. 7.

The low conductivity of hot gases (even after seeding), compared
to metals such as copper, makes it necessary to use much more powerful magnets in an MHD generator than those used in conventional
generators. Large, powerful, superconducting magnets have been built
for use in high-energy physics applications, and the technology is
expected to be applicable to MHD generator construction.

The use of MHD generators in combination with conventional
steam turbine systems appears to hold the most promise for large-scale
MHD power generation in the near future. The hot exhaust from the
MHD "topping unit" would provide heated air for the MHD combustion chamber and steam for the turbine generator. The overall efficiency of such combined systems is projected to be 50% to 60%, which
will provide fuel savings of 20% to 30% over conventional fossil-fueled,
steam turbine plants. The general use of MHD topping units by the
mid-1980s could effectively extend fossil fuel reserves. A representation
of a combined MHD-steam turbine system is shown in Figure 9—31.

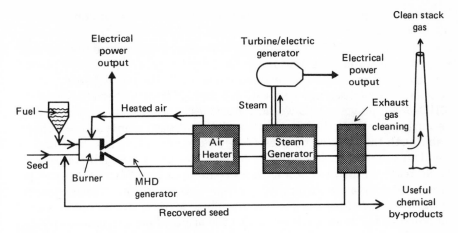

Figure 9—31. Combined MHD-Steam Turbine Power Generation

Modified from "The 1970 National Power Survey," Part 1, Federal Power Commission, Dec. 1970, p. 9—3.

Three basic approaches to MHD generator design are now being explored. These approaches are broadly classified as open cycle, closed cycle, and liquid metal systems. Our discussion to this point has dealt with the open cycle system which is closest to being commercially feasible. The closed cycle system is similar to the open cycle in that a hot, seeded gas is passed through a magnetic field. However, the working fluid, consisting of a seeded rare gas such as helium, is continuously recirculated through a closed loop. The working fluid temperatures are lower (1700°—1900°C), which creates the possibility of substituting closed cycle MHD generators for steam turbines in advanced fission generating plants. Most closed cycle research has been directed toward space and military applications; research into this new area is just beginning.

Liquid metal systems are designed to operate at still lower temperatures than closed gas systems (800°—1100°C). The liquid metal used as a working fluid has higher electrical conductivity, and high temperatures (to generate charged particles) are unnecessary. The use of liquid metal also allows lower power, nonsuperconducting magnets to be utilized. The disadvantages of liquid metal systems arise from difficulties associated with accelerating the liquid metal to velocities necessary for power generation.

Reliable cost estimates for MHD power plants are not yet available, but the general consensus seems to be that construction cost should be about the same as those for conventional power plants. The cost of magnets will be the largest single item; the air preheater is also expected to be expensive. Reductions in the costs of superconducting magnets and of oxygen (used to enrich fuel mixtures and reduce preheating requirements) would improve the prospects for commercial MHD development.

The use of MHD generators could substantially reduce the thermal pollution of water. The amount of heat discharged from a power plant is related to the energy conversion efficiency of the plant. More efficient plants discharge less heat. Because of their higher efficiencies, MHD-steam power plants would require less than half the amount of cooling water needed by conventional fossil-fueled plants and less than one fourth that needed by present nuclear power plants, as shown by the data represented in Figure 9—32. The nature of the MHD process could result in further cooling water reductions if an MHD generator is combined with a gas turbine rather than a steam turbine. No condenser coolant is needed for the gas turbine. Thus, MHD systems could be designed to reject all waste heat into the atmosphere and cause no thermal pollution of water. Also, since little cooling water is required, such plants could be located in water-poor but fuel-rich areas such as the coal-rich western United States.

A reduction in air pollution also seems to be possible by using MHD plants. Combustion takes place at very high temperatures in MHD plants, assuring the complete combustion of all fuel present. This provides a simpler mix of exhaust gas components which can be dealt with more easily than the complex mix characteristic of conventional coal-fired plants. Experiments have shown that the seeding material reacts with sulfur in the fuel to form alkali metal sulfates (such as K_2SO_4) which are relatively easily precipitated and removed from exhaust gases. Under such conditions practically all sulfur can be

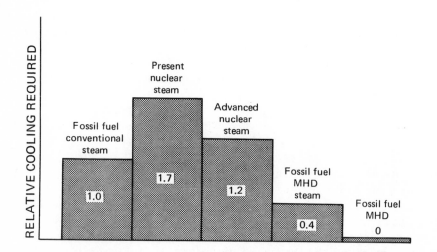

Figure 9—32.   Comparison of Required Cooling for Various Types of Power Plants

From "Energy Research and Development," House of Representatives Subcommittee Hearings, May 1972, p. 564.

removed, even that resulting from high-sulfur coals. Nitrogen oxide emissions, once expected to be a problem because of high combustion temperatures, can also be reduced. This is done by burning the coal in a fuel-rich mixture and adding excess air further downstream from the initial combustion zone. The advantages of MHD power generation seem to make it an important area for research and development in the future.

Suggested Readings

Austin, L. G., "Fuel Cells," *Scientific American*, Oct. 1959.

Barnea, J., "Geothermal Power," *Scientific American*, Jan. 1972, pp. 70–77.

Böer, K. W., "The Solar House and Its Portent," *Chemical Technology*, July 1973, pp. 394–400.

Calvin, M., "Solar Energy by Photosynthesis," *Science*, April 19, 1974, pp. 375–381.

Clancy, E. P., "Power from the Tides," reprinted in *Oceanography*, Pirie, R. G., ed., pp. 346–351, Oxford University Press, 1973.

Creager, W. P., and J. D. Justin, *Hydroelectric Handbook*, John Wiley & Sons, Inc., 1950 (Chapter 12).

Department of the Interior, "Final Environmental Statement for the Geothermal Leasing Program," Government Printing Office, Washington, D.C., 1973.

Federal Power Commission, "The 1970 National Power Survey," Government Printing Office, Washington, D.C., Dec. 1971.

"Gas-Powered Fuel Cells: Smaller and Better," *Pipeline and Gas Journal*, April 1974, pp. 46–48.

"Geothermal Energy: A National Proposal for Geothermal Resources Research," NSF/RA/N–73–003, Government Printing Office, Washington, D.C., 1973.

"Geothermal Energy Resources and Research," Hearings before the U.S. Senate Committee on Interior and Insular Affairs, June 1972, Government Printing Office, Washington, D.C.

"Geothermal Heats Up," *Environmental Science and Technology*, Aug. 1973, pp. 680–681.

"Geothermal Resources," Hearings before the U.S. Senate Subcommittee on Water and Power Resources, Aug. 1973, Government Printing Office, Washington, D.C.

Gregory, D. P., "The Hydrogen Economy," *Scientific American*, Jan. 1973, pp. 13–21.

Hammond, A. L., "Dry Geothermal Wells: Promising Experimental Results," *Science*, Oct. 5, 1973, pp. 43–44.

————, "Geothermal Energy: An Emerging Major Resource," *Science*, Sept. 15, 1972, pp. 978–980.

————, "Magnetohydrodynamic Power: More Efficient Use of Coal," *Science*, Oct. 27, 1972, pp. 386–387.

————, "Photovoltaic Cells: Direct Conversion of Solar Energy," *Science*, Nov. 17, 1972, pp. 732–733.

————, "Solar Energy: The Largest Resource," *Science*, Sept. 22, 1972, pp. 1088–1090.

Henahan, J. F., "Energy from the Earth's Core," *Saturday Review/World*, March 23, 1974, p. 64.

————, "How Soon the Sun: A Timetable for Solar Energy," *Saturday Review/World*, Nov. 20, 1973, pp. 64–66.

"Hydrogen-Economy Concept Gains Credence," *Chemical and Engineering News*, April 1, 1974, pp. 15–16.

"Hydrogen Fuel Economy: Wide-ranging Changes," *Chemical and Engineering News*, July 10, 1972, pp. 27–30.

"Hydrogen Fuel Use Calls for New Source," *Chemical and Engineering News*, July 3, 1972, pp. 16–18.

"Hydrogen: Likely Fuel of the Future," *Chemical and Engineering News,* June 26, 1972, pp. 14–17.

"Hydrogen Sought via Thermochemical Methods," *Chemical and Engineering News*, Sept. 3, 1973, pp. 32–33.

Kruger, P. and C. Otte, eds., *Geothermal Energy: Resources, Production, Stimulation*, Stanford University Press, Stanford, California, 1973.

Lear, J., "Clean Power from Inside the Earth," *Saturday Review*, Dec. 5, 1970, pp. 53–61.

Maugh, T. H., II, "Fuel Cells: Dispersed Generation of Electricity," *Science*, Dec. 22, 1972, pp. 1273–1274.

————, "Hydrogen: Synthetic Fuel of the Future," *Science*, Nov. 24, 1972, pp. 849–852.

McCaull, J., "Windmills," *Environment*, Jan./Feb. 1973, pp. 6–17.

Meinel, B. A., and M. P. Meinel, "Physics Looks at Solar Energy," *Physics Today,* Feb. 1972, p. 44.

"Methyl Fuel Could Provide Motor Fuel," *Chemical and Engineering News*, Sept. 17, 1973, pp. 23–24.

Metz, W. D., "Ocean Temperature Gradients: Solar Power from the Sea," *Science*, June 22, 1973, pp. 1266–1267.

Mills, G. A., and B. M. Harney, "Methanol – The 'New Fuel' from Coal," *Chemical Technology*, Jan. 1974, pp. 26–31.

Morrill, C. C., "Fuel Cells for the United States Energy System," in "Energy Research and Development," Hearings before the Subcommittee on Science, Research, and Development, House of Representatives, May 1972, pp. 585–593, Government Printing Office, Washington, D.C.

Morrow, W. E., "Solar Energy: Its Time Is Near," *Technology Review*, Dec. 1973, p. 31.

National Aeronautics and Space Administration, "Fuel Cells: A Survey," (NASA SP-5115), Government Printing Office, Washington, D.C.

NSF/NASA Solar Energy Panel, "Solar Energy as a National Energy Resource," NSF/RA/N–73–001, Government Printing Office, Washington, D.C.

"Proceedings of the Solar Heating and Cooling for Buildings Workshop," March 1973, NSF/RA/N–73–004, Government Printing Office, Washington, D.C.

Putnam, P. C., *Power from the Wind*, Van Nostrand Reinhold Company, New York, 1948.

Robson, G. R., "Geothermal Electricity Production," *Science*, April 19, 1974, pp. 371—375.

"Solar Heating and Cooling Demonstration Act," Hearings before the Subcommittee on Energy, House of Representatives, Government Printing Office, Washington, D.C., Nov. 1973.

Wade, N., "Windmills: The Resurrection of an Ancient Energy Technology," *Science,* June 7, 1974, pp. 1055—1059.

Wentdorf, R. H., Jr., and R. E. Hanneman, "Thermochemical Hydrogen Generation," *Science,* July 26, 1974, pp. 311—319.

Wolf, M., "Solar Energy Utilization by Physical Methods," *Science*, April 19, 1974, pp. 382—386.

Zener, C., "Solar Sea Power," *Physics Today*, Jan. 1973, pp. 48—53.

Part 3

Energy Conservation

Solar furnace, Odello, France: Courtesy Georg Gerster, Rapho Guillumette Pictures

Chapter 10

Energy Conservation

The information presented in previous chapters indicates that fossil fuels are now the main sources of usable energy in the United States. According to recent estimates, they will continue to satisfy more than 50% of the annual energy requirements of the United States at least through 1985. We have also seen that the supply of these fuels is finite and will someday be exhausted. Realistic conservation measures have been proposed which, if implemented, can significantly decrease the use of all forms of energy and push that unfortunate day farther away into the future. In this chapter we will investigate these proposals for each of the four areas of energy use.

Benefits of Energy Conservation

A number of benefits can be derived from an energy conservation program. Four important ones we will discuss are (1) conserving a valuable resource, (2) gaining time to develop alternatives to fossil fuel energy sources, (3) decreasing environmental pollution, and (4) minimizing the dependence of the United States on imported energy sources.

Figure 10—1 shows the quantity of fossil fuels consumed in 1971 and estimates of future consumption both with and without effective conservation programs. It is quite apparent that significant reductions in consumption are considered possible. These resources should be conserved as much as possible, not only for their energy content but also because of their value as sources of chemicals. Some scientists have even advocated that coal should not be used as a fuel at all, since it represents the main source of nearly all of the aromatic chemicals that are used to produce many important substances such as plastics, drugs, other medicines, vitamins, and dyes. It is not likely that such recommendations will be followed, but fossil fuel conservation will extend the useful life of this valuable resource.

As we have seen in previous chapters, energy sources other than fossil fuels are now being used and others, especially fast-breeder reactors and nuclear fusion sources, are being developed. It is the opinion of experts that 10—25 years will be needed to perfect the

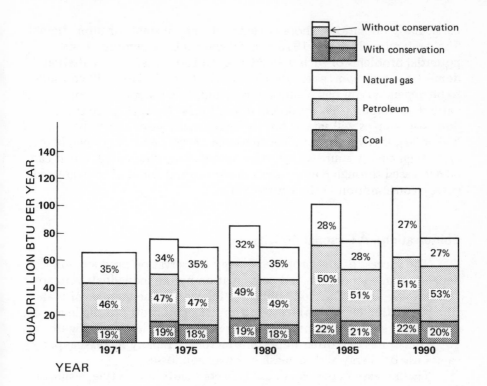

Figure 10—1. United States Fossil Fuel Consumption with and without Suggested
Conservation Programs

From "The Potential for Energy Conservation: A Staff Study," Executive Office of the
President, Office of Emergency Preparedness, Oct. 1972, p. 50.

technologies and build the facilities required to make the widespread use
of such energy possible. The conservation of fossil fuels allows more
time for such developmental activities.

It has also been pointed out previously that a significant problem
involved in the developing energy crisis is the conflict between environmental protection and energy production. The only completely
pollution-free energy source is sunlight, and, as we have seen, the direct
use of sunlight or its conversion into other energy forms (electricity) is
not likely to contribute significantly to satisfying the energy demands
of the nation. It follows, then, that a decrease in use of the other
(polluting) forms of energy will lead to decreases in air pollution
(particulates, sulfur oxides, nitrogen oxides, carbon monoxide, and
hydrocarbons), water pollution (heat and oil spills), and the problems
of radioactive waste disposal.

In Chapter 5 we discussed a number of problems relating to
petroleum imports. These problems indicate how unwise it is for a
nation to depend heavily on other nations for essential energy supplies —

particularly on nations whose governments are unstable or unpredictable. The Arab oil embargo of 1973—74 provided a brief glimpse of the potential problems of such dependence. Unfortunately, until alternate domestic energy sources are developed, the United States will continue to be dependent on petroleum imports and, to a lesser extent, on natural gas imports. This problem doesn't exist for coal, since the domestic sources will satisfy our needs for many years (possibly 300—500), if environmental conflicts can be resolved. Our dependence on foreign energy sources cannot be immediately eliminated, but it can be decreased through conservation measures that will lead to decreased energy consumption in the United States.

Conservation: A Personal Activity

The success of any proposed energy conservation program will depend upon cooperation and participation by the individuals who are the ultimate users. The well-publicized increase in the number of users (the population explosion) has contributed significantly to the increasing demand for energy, but more than mere numbers is involved. Figure 10—2 illustrates two other important factors: a rapidly increasing per capita use of energy and increasing conversion losses.

The increase in per capita use reflects a shift in life-style, a substitution of motor power for muscle power. This substitution is often

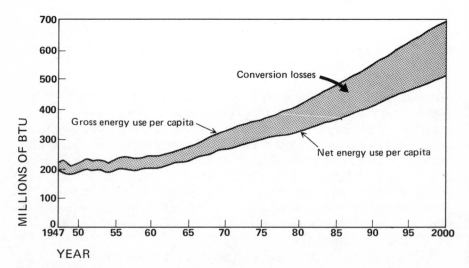

Figure 10—2.   Per Capita Gross and Net Energy Use in the United States, 1947—2000

From "United States Energy Through the Year 2000," U.S. Dept. of the Interior, Dec. 1972, p. 5.

questionable, as in the use of motor-driven lawn mowers, can openers, and tooth brushes. We are living in a convenience-oriented society.

The increasing difference between gross (input) energy and net (used) energy is the result of a shift in energy use from primary sources such as coal, natural gas, and oil to secondary sources, mainly electricity. The energy losses encountered when one form is changed into another were discussed in Chapter 2. These unavoidable losses make it necessary to consume more fuel to provide one Btu of useful electrical energy than would be necessary if the used energy were obtained directly from the burning fuel. Thus, the shift to the use of electrical energy causes a larger than equivalent increase in the consumption of primary fuels.

How can the participation and cooperation of the individual energy user be obtained? Participation can obviously be forced by the passage of appropriate laws. Suggestions have been made, for example, that laws be passed to accomplish the following:

1. Ration energy use
2. Tax excessive energy use
3. Limit the number of automobiles per family
4. Limit the number of children per family
5. Curtail immigration into the United States
6. Refuse foreign aid to countries lacking birth control laws
7. Specifically limit energy use by upper and middle economic classes
8. Make electric heating of homes illegal if natural gas is available
9. Make it illegal to use natural gas for the generation of electricity

The enactment of some of these laws or modifications of them might be necessary and even desirable, but, hopefully, individual cooperation can be obtained in other ways. A massive government-financed public relations campaign involving all the mass media has been proposed. The campaign would be somewhat similar to the well-known efforts directed against pollution and littering. It is likely that the large commercial advertisers would cooperate by extolling the favorable energy characteristics of their products and providing added momentum for the project.

Regardless of the methods used to obtain participation, an effective energy conservation program will require changes to a simpler life-style for most of us. Some specific proposals for conservation are presented later in this chapter, but in general the net effect will be a decreased per capita use of goods and energy. Some of these decreases will be accomplished by seemingly minor activities such as turning off unnecessary lights, walking rather than driving for short trips, and washing dishes or clothes in warm rather than hot water. Some typical major changes involve decreases in the size, power and number of automobiles, the increased development and use of mass transit systems, and a decrease in the length of vacation trips.

Energy conservation can be accomplished in two ways. The first (belt tightening) is to use less energy and in the process sacrifice some of the goods and conveniences associated with its use. The second approach (leak plugging) is to use energy more efficiently but maintain the level of consumption of goods and the use of conveniences. Both approaches are represented in the conservation proposals later in this chapter.

Another useful way of classifying energy conservation proposals is according to the time required for the conservation activities to be initiated and have an effect on energy use. On this basis, short-term proposals require 1 to 3 years and include many practices that can be carried out on a personal, individual basis. Midterm proposals require 4 to 8 years to be effective and in most cases make use of known technology. However, individual action is limited because changes in such things as the design of consumer products are involved. Long-term proposals require more than 8 years for implementation and often necessitate the development of new technologies or the incorporation of significant changes in established practices (such as modes of transportation or the design of urban communities).

Future Energy Demand and Supply

An aid to understanding the potential for energy savings through conservation is provided by an investigation of energy demand-and-supply projections for the future. In Figure 10—3 the energy demand by use area is given as projected through 1990. Notice that the energy is given in units of quadrillion (10^{15}) Btu, where one quadrillion Btu is approximately equal to the energy in 172 million barrels of oil, 970

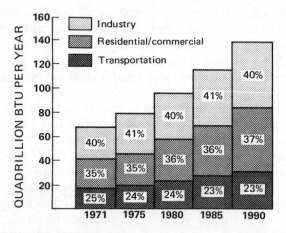

Figure 10—3.   United States Energy Demand According to Use Areas

Adapted from "The Potential for Energy Conservation: A Staff Study," Executive Office of the President, Office of Emergency Preparedness, Oct. 1972, Fig. II-2, p. 7.

billion cubic feet of natural gas, or 42 million tons of coal. The total energy demand is predicted to double in the twenty years covered by the projections, but the percentage of the total in each use area remains essentially constant. The residential and commercial areas are considered together because conservation proposals are similar for each of them.

Estimates of the future major sources of energy in the United States were given earlier (see Figures 1—3 and 1—4). It is apparent that fossil fuels are now the primary sources of energy; they are projected to retain this role into the foreseeable future. The percentages of the total energy derived from hydroelectric sources, coal, and petroleum are expected to remain essentially constant through 1990. The contribution of natural gas to the total is expected to decrease as nuclear energy sources become increasingly important.

The projections represented by Figures 1—3, 1—4, and 10—3 were made on the assumption that energy use patterns will change very little in the United States between now and 1990. However, conservation programs can have a significant impact on energy use, as shown by the projections given in Figure 10—4. According to these estimates, the annual use of energy in the United States can be reduced by more than

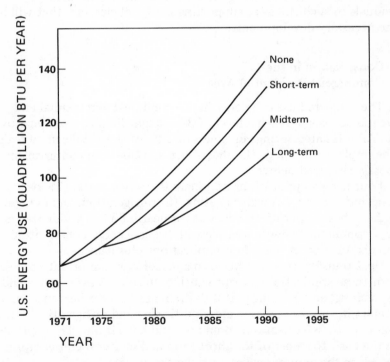

Figure 10—4.   Idealized Energy Use Projections Based on Suggested Conservation Measures

From "The Potential for Energy Conservation: A Staff Study," Executive Office of the President, Office of Emergency Preparedness, Oct. 1972, p. 59.

24% by 1990 if short-term, midterm, and long-term conservation measures are implemented.

Energy Conservation in Use Areas

The remainder of this chapter is devoted to an investigation of specific energy conservation recommendations for the use areas, commercial-residential, transportation, and industrial. The commercial and residential areas are treated together since many of the primary energy uses (and therefore the conservation measures) are the same.

Energy conservation for each area will be looked at from the point of view of the short-term, midterm, and long-term proposals. As mentioned earlier, short-term measures often include activities that can easily be undertaken by individual energy users, while midterm measures often require improvements in consumer products or services. Individuals can become directly involved in implementing midterm measures by purchasing and using only the most energy-efficient goods or services, and by indicating, through appropriate activities, that such goods or services are desired. Long-term proposals requiring the development of new technologies and procedures or the extensive improvement of current ones generally limit the involvement of individuals to such things as supporting laws, policies, etc. that will lead to the necessary developments.

Conservation in the
Commercial-Residential Area

The projected use of energy in the combined commercial and residential areas will reach a level of 34.6 quadrillion Btu per year by 1980. An estimated savings in excess of 14% of this total can be realized by the implementation of the short-term and midterm conservation proposals discussed below.

Four energy applications dominate these two areas: (1) space heating and cooling, (2) water heating, (3) refrigeration, and (4) cooking. Together, these applications account for nearly 88% of residential and 77% of commercial energy use. We will discuss energy conservation proposals for each of these four applications plus lighting.

Heat transfer from one place to another is an important process in each of these applications except lighting. In some cases — space and water heating, and cooking — it is desirable to transfer heat into an area and retain it. In other cases — air conditioning and refrigeration — it is desirable to transfer heat out of an area and keep it out. These processes form the basis for most of the short-term and midterm conservation proposals in the commercial and residential areas.

Remember that heat can be transferred from one place to another by three different processes known as conduction, convection, and radiation. Conduction is the process involved when heat travels through a substance without causing any visible motion of the substance. For example, a metal poker with one end held in a fire soon becomes hot at the other end. The heat always flows from a hotter area to a cooler area. The amount of heat conducted through a substance per hour is given by the following equation:

$$H = \frac{KA\Delta t}{\ell}$$

10–1

In this equation, H is the number of Btu conducted in one hour, K is a constant called the thermal conductivity of the conducting substance, A is the area of the conducting substance in square feet, ℓ is the thickness of the conducting substance in inches, and Δt is the difference in temperature (°F) between the two sides of the conducting substance. The following example illustrates the use of this equation.

Example 10–1. How much heat is lost per hour through a brick wall of a room if the area of the wall is 80 square feet, the thickness is 3 inches, the room temperature is 72°F, and the outside temperature is 32°F? The thermal conductivity of brick is 4.35.

$$H = \frac{KA\Delta t}{\ell} = \frac{(4.35)(80)(40)}{(3)} = 4637 \text{ Btu per hour}$$

The thermal conductivities of various substances are given in Table 10–1.

We see from equation 10–1 that heat flow through a substance is enhanced by large areas (A), thin walls (ℓ), large temperature differences (Δt), and high thermal conductivities (K). Thus large, thin-walled, metallic surfaces make good heat exchangers. To prevent heat flow,

Table 10–1. THERMAL CONDUCTIVITIES

Material	Thermal Conductivity
Air at rest	0.165
Aluminum	1393
Asbestos paper	1.74
Brass	755
Brick	4.35
Copper	2636
Rock wool	0.258
Steel	334
Window glass	7.26
Wood (fir)	0.261

thick layers of substances having low thermal conductivities are used. This is the basis of insulation.

On the basis of the low thermal conductivity for air listed in Table 10—1, we might imagine that a blanket of air could serve as a good insulator. However, if we attempt to use air for this purpose, we find that it transfers heat quite well in spite of the low thermal conductivity. The heat is transferred by convection, which is the transport of heat by the motion of the matter containing it. Air in contact with a warm surface is heated by conduction. The heated air expands and becomes less dense than the surrounding cooler air. The warm, lower density air rises and cooler air moves in to take its place. This process creates currents of air which circulate the warm air away from the warm surface and thus transfer heat. The process is represented in Figure 10—5.

The third process of heat transfer, radiation, is employed when we warm our hands by holding them in front of an open fire, even though the surrounding air is cold. It is also the means by which the energy of the sun travels through space to the earth. It is a process of heat transfer that does not require any transferring substance. Radiant energy is sometimes visible to the eye, as in the case of a glowing light bulb. However, radiant energy is also given off as heat. A warm tea kettle is an example. Each of these heat transfer processes can be important factors in energy conservation attempts.

It is estimated that by 1980 the annual savings in energy for the commercial-residential area will be 500 trillion Btu in cooling and 2 quadrillion Btu in heating if the short-term and midterm measures given below are followed. This represents a reduction of more than 22%.

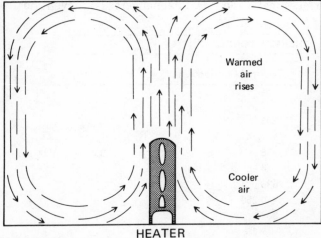

HEATER

Figure 10—5.   Transfer of Heat by Convection

Short-term measures:

1. Use blinds and drapes effectively. Close them in cold weather to prevent interior heat loss by radiation and convection. Close them in warm weather to keep out exterior heat. If direct sunlight falls on windows, open drapes or blinds in cold weather to let in radiant heat.
2. Adjust the heating and cooling thermostat carefully. Set it as low as is comfortable in cold weather and as high as is acceptable in warm weather. Just be warm, not hot, in the winter and cool, not cold, in the summer.
3. Clean the furnace and air-filtering system regularly to allow easy transfer of heated or cooled air.
4. Keep condenser coils (heat exchangers) of air conditioners clean to allow efficient conductive heat exchange to occur.
5. Use light-colored roofing and outside paint on buildings. This will reflect large amounts of incoming radiant energy and aid in the cooling process.
6. Use awnings or natural shade to help keep buildings cool.

Midterm measures:

1. Develop more efficient furnaces. It has been suggested that electric ignition be used in gas furnaces instead of the wasteful burning pilot light.
2. Use solar heating where possible. Fuel costs will influence the feasibility of this proposal; as fuel costs go up, the proposal becomes more feasible.
3. Make wide use of shade trees. This could become a long-term proposal unless fast-growing trees are used.
4. Use "superinsulation" in buildings. It has been estimated that 40%–70% of the heating and cooling energy of buildings (mainly homes) could be saved if superinsulation were used. This approach involves the use of maximum amounts of insulating materials in roofs, walls, and under floors; double-glazed glass in windows; storm doors, and careful weather stripping around doors and windows. Ordinary and superinsulation are compared in Figure 10—6.

Long-term measures:

1. Develop the district-heating concept: use excess or waste steam from centralized power plants to provide heat for surrounding commercial and residential customers. In order to make this proposal feasible, fuel costs would have to increase and new ideas would have to be incorporated into the spatial relationships between communities and power plants (communities would be built near and around power plants rather than away from them).

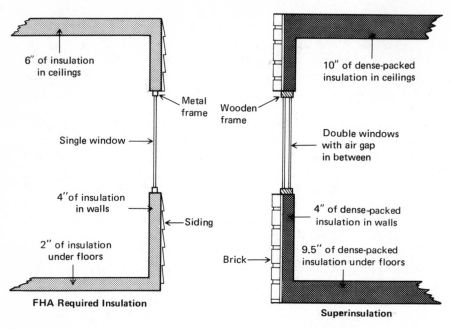

FHA Required Insulation

6″ of insulation in ceilings

Metal frame

Wooden frame

Single window

4″of insulation in walls

Siding

2″ of insulation under floors

Superinsulation

10″ of dense-packed insulation in ceilings

Double windows with air gap in between

4″ of dense-packed insulation in walls

Brick

9.5″ of dense-packed insulation under floors

Figure 10—6.   Ordinary Insulation and Superinsulation

2.   Include municipal wastes as a part of the fuel used in power plants. This also aids in the problem of solid waste disposal.

The adoption of short-term and midterm conservation measures in water heating could result in annual energy savings of 2.5 quadrillion Btu by 1980 in the residential area alone. A more realistic figure is about 10% of this potential or 250 trillion Btu.

Short-term measures:

1.   Keep heat transfer surfaces in water heaters clean. This allows efficient conductive heat transfer into the water.
2.   Use only full loads in clothes washers and dishwashers. Partial loads use as much hot water as full loads.
3.   Wash clothes and dishes in warm or even cold water when possible.
4.   Repair and maintain all leaky hot water faucets.

Midterm measures:

1.   Increase insulation of heater shell. This decreases conductive heat losses to the surroundings.
2.   Insulate hot water pipes. This also decreases conductive heat losses to the surroundings.

3. Develop more efficient burners. This would enable a more complete combustion of fuel with a correspondingly greater release of heat energy.

4. Use exhaust gases as a space heating supplement. Hot exhaust gases could be passed through heat exchangers located in living areas. Some heat will be given up by the hot exhaust to the living area. See Figure 10—7.

5. Recover heat from used hot water. By means of an appropriate arrangement of drain and inlet pipes, some heat from used hot water could be transferred to incoming water. See Figure 10—7.

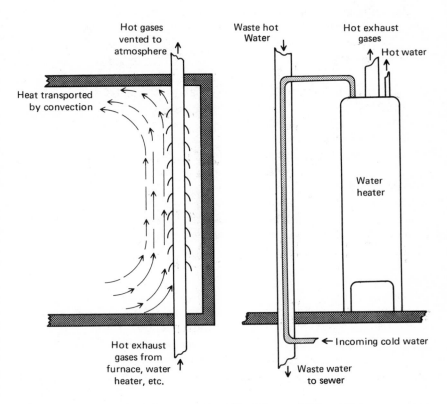

Figure 10—7.   The Extraction of Heat from Used Water and Exhaust Gases

Long-term measures:

1. Use solar energy to heat water. The feasibility of this proposal depends on fuel costs and climate. It is used in some areas of the world, and water temperatures of up to 200° F are obtained.

2. Use excess or waste steam from power plants to heat water. This is the district-heating concept again.

It is estimated that by 1980 100 trillion Btu per year can be saved if the following short-term and midterm measures are applied to refrigeration systems.

Short-term measures:

1. Keep condenser coils clean. This allows efficient conductive heat exchange to take place.
2. Replace old gasket around door to insure a tight seal.
3. Avoid opening unnecessarily. Open as seldom as possible. Get all necessary items during one opening. Use externally located cold water and ice sources, if possible. Encourage manufacturers to phase out models that do not have such exterior sources.

Midterm measures:

1. Increase box insulation thickness. This minimizes the conductive flow of heat into the box from the surroundings.
2. Compartmentalize the interior. Only a part of the interior would need to be opened at any time.
3. Develop more efficient electrical motors. More friction-free bearings would decrease energy losses to the surroundings during motor operation.

The implementation of both short-term and midterm conservation proposals will result in a savings of 58 trillion Btu per year by 1980 in residential cooking use alone.

Short-term measures:

1. Carefully use electric stoves. Turn off heat before cooking is complete and let the residual heat finish the job.
2. Use the lowest acceptable settings on the stove. This minimizes the heat lost to the surroundings from around pans, etc.
3. Avoid the use of highly specialized gadgets. Bacon friers, for example.

Midterm measures:

1. Insulate sides and covers of pans. This will prevent heat loss by conduction to the surroundings.
2. Reshape pots and pans to direct the heat to where it is wanted.
3. Replace gas pilot lights with electrical igniters.
4. Insulate ovens better. This is especially useful for cutting down on air conditioning requirements.
5. Establish minimum efficiency standards for appliances and mark the efficiency of each appliance on the name plate. This will allow the consumer to shop for the most energy efficient appliances.

Energy for lighting in the commercial-residential area accounts for about 1.5% of the national total. Yet out of this relatively small percentage, a savings of 350 trillion Btu per year is projected for 1980 with the adoption of the short-term and midterm measures given below.

Short-term measures:

1. Turn off lights upon leaving a room.
2. Use lower light intensities in rooms. This can be done now by the use of lower wattage light bulbs. Lower intensities can be designed into buildings in the future (midterm).
3. Use natural light whenever possible.
4. Use daylight savings time on a permanent basis. This would presumably provide natural light for use during more of the active part of each day, but experience has led to differences of opinion concerning the actual energy savings.

Midterm measures:

1. Develop automatic sensors to turn off lights after last occupant has left a room.
2. Develop the efficient fluorescent light for use in the available screw-in type socket. Fluorescent lights are three times as efficient as the incandescent variety.
3. Design lower light intensities into the living and working spaces of new or remodeled buildings. Measurements show that this can be done without harming the eyes or impairing the working efficiency of people.
4. Reduce light intensities for parking lots, streets, and highways where safety permits.
5. Avoid (make illegal?) the excessive use of lights in advertising.
6. Avoid picturesque but very inefficient gas lighting.

The total annual energy saving estimated to be possible by 1980 in the commercial-residential area is summarized in Table 10—2.

Conservation in the Transportation Area

In 1970 the transportation area accounted for the use of 16.4 quadrillion Btu of energy — about 25% of the national total. Predictions are that approximately the same percentage of the national total will be used for transportation in the future. Petroleum was the source for 96% of the energy used for transportation in 1970. Coal provided only a negligible part, and natural gas was mainly used to provide pumping power for moving the gas itself through pipelines. Gasoline was the dominant petroleum product used, as indicated by

Table 10—2.   ESTIMATED ANNUAL ENERGY SAVINGS POSSIBLE IN THE
COMMERCIAL-RESIDENTIAL AREA, 1980

	Potential Savings (trillions of Btu)	Percent of Area Total Saved	Percent of National Total Saved
Residential Area Uses			
Space heating and cooling			
existing homes	1100	3.1	1.1
new homes	1100	3.1	1.1
Water heating	250	0.7	0.3
Cooking	50	0.1	0.05
Refrigeration	100	0.3	0.1
Air conditioning equipment	500	1.4	0.5
Other, including lighting,			
clothes drying, etc.	500	1.4	0.5
Residential total	3600	10.1	3.7
Commercial Area Uses			
All commercial uses	1500	4.2	1.6
Commercial total	1500	4.2	1.6
Combined total	5100	14.3	5.3

From "The Potential for Energy Conservation: A Staff Study," Executive Office of the President, Office of Emergency Preparedness, Oct. 1972, p. D—12.

the fact that automobiles accounted for more than 55% of the petroleum consumption in the area. Trucks used 21% of the total, aircraft 7.5% (up from 4% in 1960), and railroads 3.3% (down from 4.0% in 1960).

During the last few decades dramatic shifts have taken place in the modes of transportation used. The general trend has been away from the more energy-efficient (but slower and less convenient) modes, toward the less energy-efficient (but faster and more convenient) modes. This has occurred in both the transporting of people, where airplanes and automobiles have displaced buses and railroads, and the transporting of freight, where the use of trucks and airplanes is increasing at the expense of pipelines, waterways, and railroads. These trends are shown graphically in Figures 10—8 and 10—9. Note the separate scale used for airlines in Figure 10—9.

The effects of using less efficient transportation modes become more apparent when the efficiencies of the various modes are

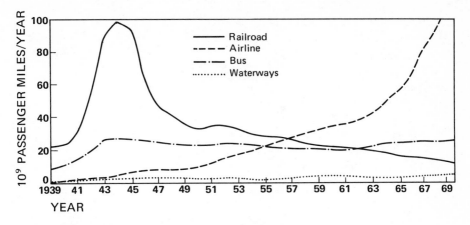

Figure 10—8.   Intercity Transportation by Public Carriers

Adapted with permission from *Transportation Facts and Trends,* April 1971, copyright by the Transportation Association of America.

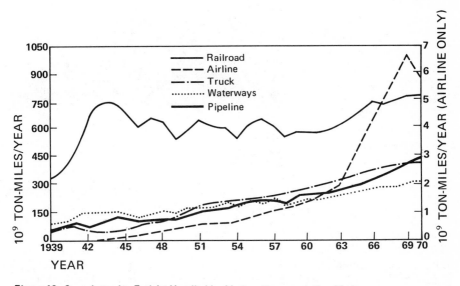

Figure 10—9.   Intercity Freight Handled by Various Transportation Modes

Adapted with permission from *Transportation Facts and Trends,* April 1971, copyright by the Transportation Association of America. Airline data selected from *Handbook of Airline Statistics,* 1971 edition, Civil Aeronautics Board, pp. 12, 17.

compared. This is done in Figure 10—10. Note the number of passengers on which the efficiency is calculated. These numbers are not maximum capacities but average figures based on present-day experience. Efficiencies can be increased dramatically by adding more passengers.

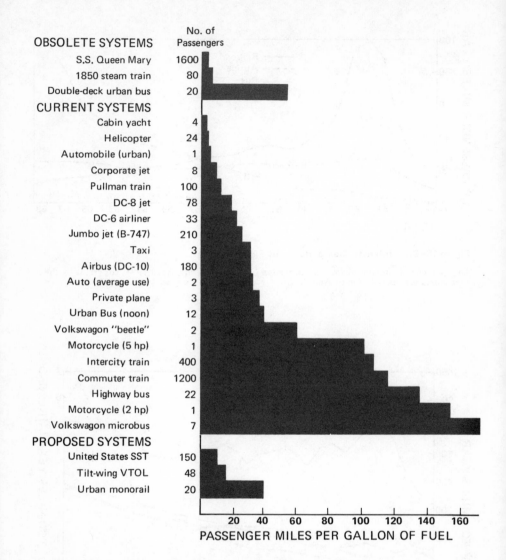

Figure 10—10.   **Efficiencies of Various Transportation Modes, 1972**

Redrawn from "System Energy and Future Transportation" by Richard A. Rice, in *Technology Review* (January 1972) by permission of the Alumni Association, Massachusetts Institute of Technology.

A shift to more energy-efficient modes of transportation is an often-proposed conservation measure. The validity of such action is illustrated by the data of Table 10—3.

Figure 10—11 shows projections of petroleum use in the transportation area through 1990; the impact of the automobile is obvious. Because of this dominant role, many of the more widely publicized energy conservation measures have been aimed at reducing automobile

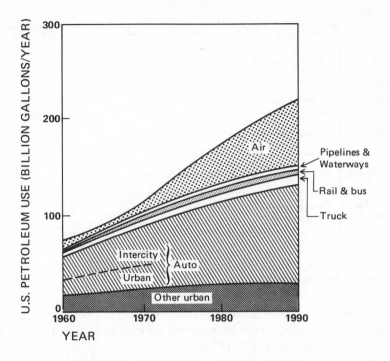

Figure 10—11. United States Use of Petroleum in Transportation

Modified with permission from *Transportation Energy and Environmental Issues* by W. E. Fraize and J. K. Dukowicz, Mitre Corporation, Feb. 1972. Figure is based on R. A. Rice, *System Energy as a Factor in Considering Future Transportation*, ASME Paper 70—WA/ENER—8, Nov. 1970.

Table 10—3. FUEL CONSERVATION RESULTING FROM VARIOUS CHANGES IN TRANSPORTATION MODES

Proposed Change	Fuel Savings (%)
50% of present urban commuters form car pools	3.1
50% of present commuters (to and from city centers) use bus service	1.9
50% of present intercity auto passengers use bus and rail service	3.0
50% of present intercity trucked freight switched to rail freight	3.4
50% of present short-haul air passengers use intercity bus service	0.3

From "Conservation and Efficient Use of Energy — Part 3," U.S. House of Representatives Subcommittee Hearings, July 11, 1973, p. 1495.

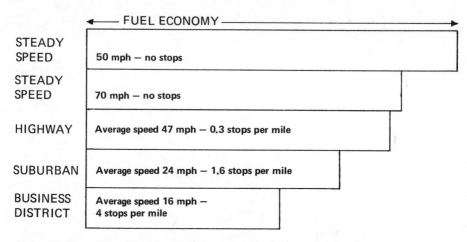

Figure 10—12.   Effect of Type of Driving on Fuel Economy (level road)

From "Conservation and Efficient Use of Energy — Part 3," U.S. House of Representatives Subcommittee Hearings, July 11, 1973, p. 1151.

fuel consumption. These range from suggestions for decreasing automobile use and weight to the nationwide 55-mile-per-hour speed limit which became law January 2, 1974.

Fuel economy for vehicles is usually represented by the miles traveled per gallon of fuel (mpg), and for light-duty vehicles varies between 5 and 50 mpg. This economy is influenced by the following major factors: (1) the type of route traveled, (2) the manner in which the vehicle is driven, (3) the design of the vehicle, (4) the engine design, (5) the type and number of accessories and (6) the driving conditions.

Figure 10—12 illustrates the effects of various types of driving on fuel economy. No mpg figures are used since the figure is intended only to illustrate the relative effects. The dependence of fuel economy on both speed and number of stops per mile is quite apparent. Conservation efforts directed at changing the type of driving have been focused primarily on reversing two trends: increased use of higher speeds on highways and short-distance (stop-and-go) use of automobiles.

The somewhat controversial nationwide 55-mph speed limit is based on sound scientific information — driving at lower speeds generally results in substantial fuel savings. The amount of fuel saved increases when the same speed reduction (10 mph, for example) is applied at higher speeds, as shown by the data of Table 10—4. These data are for 1965—68 model autos, but current models save similar percentages of fuel even though they get fewer miles per gallon at all speeds. The total fuel savings resulting from nationwide highway speed reductions does not approach the values given in Table 10—4 because most travel is not done at highway speeds. The reduction to 55 mph on highways has the potential to save between 1.5 and 6% of the motor vehicle fuel used annually in the U.S.

Table 10—4. FUEL SAVINGS WITH SPEED REDUCTIONS (1965—68 MODEL AUTOS)

Speed Reduction (mph)	Fuel Savings (%) Range of all autos tested	Typical auto
70 to 60	17 to 24	20
60 to 50	11 to 18	15
50 to 40	6 to 11	9

From "Energy Emergency Legislation, Part 2," U.S. Senate Subcommittee Hearings, Nov. 1973, p. 566.

Automobiles are being used increasingly for short-distance transportation — a use characterized by poor fuel economy. As shown in Figure 10—13, 54% of all automobile trips are less than 5 miles in length. This statistic indicates that other modes of transportation — walking, bicycling, etc. — could be used for a significant number of trips.

The manner in which a vehicle is driven dramatically affects fuel economy. In general, automobiles driven "hard" show higher fuel consumption than identical models driven carefully. Hard driving is characterized by such things as jack rabbit starts, rapid acceleration, high cruising speeds, uneven speeds, and engine racing when the car is

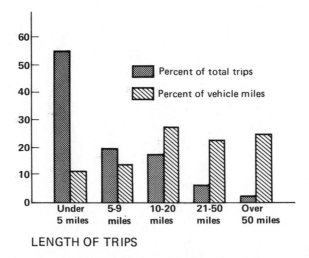

LENGTH OF TRIPS

Figure 10—13. Distribution of Automobile Trips by Distance, 1970

From "The Potential for Energy Conservation: A Staff Study," Executive Office of the President, Office of Emergency Preparedness, Oct. 1972, Fig. C—6, top of p. C—9.

not moving. The effects of such practices are known to be significant, but analyzed data are not available to illustrate this point.

Factors of automobile design most often associated with fuel economy are weight, rolling resistance, and axle ratio. Of these, weight is the single most important factor — a 5000-pound vehicle will get about 50% lower fuel economy than a 2500-pound vehicle.

Engine design, adjustment, state of tune, and overall condition significantly affect fuel economy. Periodic (and regular) tune-ups are essential if maximum fuel economy is to be realized. One misfiring spark plug can decrease fuel economy by as much as 8% in a full-sized automobile.

Optional equipment and accessories also influence fuel economy. Power brakes and steering cause practically no measurable decrease in fuel economy, but the same cannot be said about air conditioners and automatic transmissions. Tests conducted by the EPA on full-sized cars equipped with air conditioning showed a loss of 9% in fuel economy when the air conditioners were operated (ambient temperature of 70° F). This fuel penalty increased to as much as 20% during continuous air conditioner use in urban traffic at higher ambient temperatures. Similar tests revealed a fuel penalty of 5—6% for automatic transmissions.

One type of automotive equipment — the federally required emission control devices — reveal an apparent conflict between two governmental programs. The program to combat air pollution requires the installation of antipollution equipment on all automobiles. However, most of the required equipment decreases the efficiency of fuel use, as shown in Table 10—5. The average loss in fuel economy attributable to

Table 10—5.   EFFECT OF EMISSION CONTROL DEVICES
ON AUTOMOBILE FUEL ECONOMY

Model Year	Fuel Economy Loss (% of uncontrolled)
1957—67 (uncontrolled, no devices installed)	—
1968	9.6
1969	7.9
1970	7.3
1971	9.8
1972	5.3
1973	6.6
Average loss (1968—73)	7.75

From "Conservation and Efficient Use of Energy — Part 1," U.S. House of Representatives Subcommittee Hearings, June 19, 1973, p. 141.

antipollution equipment for the period 1968—73 (7.7%) is roughly
equal to that resulting from the use of convenience devices such as air
conditioning and automatic transmissions.

Energy conservation in the transportation area will be difficult
to accomplish because it will require changes in the tastes, habits,
and aspirations of the public. A number of the conservation proposals
contain suggestions that might be interpreted as threats to the speed,
comfort, convenience, and freedom now enjoyed by the traveling
citizen. In spite of the problems, it seems certain that conservation in
this area must be accomplished. Projected growth patterns indicate
an unacceptable annual consumption level of 29.2 quadrillion Btu
by aircraft and automobiles alone in the year 2000. This can be
reduced by 6.1 quadrillion Btu (20%) by the implementation of a
steady (but nonrevolutionary) shift toward more efficient modes of
transportation.

Short-term measures:

1. Increase occupancy ratio of automobiles. Establish car pools, etc.
2. Adopt energy-efficient driving habits. Use low idling rates for
    engines. Avoid rapid acceleration. Travel at lower speeds.
3. Substitute communication for travel. Phone or write rather
    than visit in person.
4. Use nearby rather than distant leisure-time facilities.
5. Use self-propulsion modes as much as possible. Walk and ride
    bicycles. Encourage the development of walkways and bike
    routes.
6. Use smaller automobiles.
7. Use low loss tires (radial). Fuel savings of up to 15% have been
    reported as a result of using the easier-rolling, radial tires.
8. Use existing mass transit systems as much as possible. Support
    efforts to upgrade and maintain railroads, bus lines, subways,
    etc.

Midterm measures:

1. Improve traffic flow. Suggested methods include appropriate
    zoning, improved intersection design, and computer-controlled
    traffic signals.
2. Develop more efficient engines.
3. Develop new mass transit systems.
4. Discourage (by taxation) the use of inefficient transportation
    modes. Prevent the use of long-distance air travel rates as a
    subsidy for short-trip rates (the reverse has even been suggested).
5. Make second cars prohibitive by increased purchase costs or
    taxation.

6.   Improve existing mass transit systems in both facilities and services. Make travel by rail once again an enjoyable experience.
7.   Consolidate freight movement in urban areas. Don't use two trucks when one will do. Use containerization to improve the efficiency of freight handling.
8.   Eliminate sales propaganda geared to the "more-, bigger-, faster-is-better" approach.

Long-term measures:

1.   Continue to develop and improve mass transit systems.
2.   Develop more efficient engines or other propulsion systems.
3.   Develop new efficient ways to transport people and things.
4.   Shift from petroleum to coal as a fuel. This will require the development of electric-powered propulsion systems.
5.   Change urban design. Cluster homes and services to allow for the widespread use of self-propulsion modes.

Conservation in the Industrial Area

The industrial area annually accounts for nearly 42% of the total national energy use. Three activities account for more than 50% of the industrial energy use: the production of primary metals (iron, steel, aluminum, copper, etc.), the production of chemicals and allied products (sulfuric acid, ammonia, plastics, etc.), and the refining of petroleum to give gasoline, motor oil, etc.

In contrast to the transportation area, energy use efficiency in the industrial area has improved markedly over the last few years. This is illustrated for the steelmaking industry by the data of Table 10–6. The improvement in energy efficiency for steelmaking is seen to parallel an increased use of the basic oxygen process. This process represents a new technology in which oxygen gas is used to quickly convert molten iron into steel. The older techniques (mainly open hearth and electric furnace) made it necessary to keep the iron molten for a longer time at great energy expense. The basic oxygen process is an example of using new technologies to conserve energy. It is estimated that the adoption of the basic oxygen process throughout the entire steelmaking industry can lead to an ultimate energy use reduction of 50% for the industry or about 3% of the annual national requirement.

.The aluminum-producing industry has demonstrated similar efficiency improvements. Aluminum is produced by passing a direct electric current through a molten mixture containing aluminum oxide (bauxite). Figure 10–14 shows how the electrical energy requirement per ton of produced aluminum decreased in the years from 1947 to

1967. In 1958 the demand for aluminum was low, and only the most efficient plants were used. The high efficiency shown by these plants gives an indication of the potential energy efficiency for the industry.

It has been estimated that, with the possible exception of the already efficient primary metals production, industrial activities can be modified to reduce energy use by a minimum of 10–15%. The general steps required are to retire older inefficient equipment and replace it with more efficient models, to redesign processes to make them more efficient, and to improve maintenance of existing equipment to maintain efficiencies as high as possible.

Table 10–6.   ENERGY USE EFFICIENCY IN THE STEEL INDUSTRY

| Year | Steel Output by Process (millions of net tons) | | | | | Energy Used Per Ton of Raw Steel Produced (millions of Btu) |
	Basic Oxygen	Electric Furnace	Open Hearth	Bessemer Converter	Total	
1960	3	8	86	1	98	30.0
1962	6	9	83	1	99	28.5
1964	15	13	98	1	127	26.6
1966	34	15	85	–	134	26.0
1968	49	17	66	–	132	26.0

Data selected from "The Potential for Energy Conservation: A Staff Study," Executive Office of the President, Office of Emergency Preparedness, Oct. 1972, Table E–3, p. E–9.

The implementation of these general steps might require governmental as well as consumer activity. A number of the proposed measures given below reflect this fact. These measures are not classified as short-term, midterm, or long-term because many of them could fit into any one of the categories depending upon the pressure exerted by the federal government, other regulating agencies, or the public.

Conservation measures for the industrial area:

1. Increase the price of energy. Charge more for large quantities rather than less as is done today. This would encourage the development of more efficient processes.
2. Tax energy use. Make energy too expensive to waste.
3. Recycle materials. The use of recycled nonferrous metals requires less than 20% of the energy needed to produce the metals from ores. It is estimated that at least 22% of the energy needed to produce a new automobile could be saved by optimum

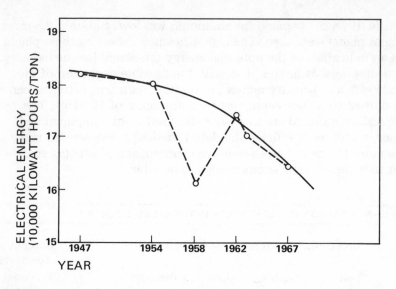

Figure 10—14.   Energy Use Efficiency in the Aluminum Industry

Modified with permission from *Interim Report: The Growing Demand for Energy,* April 1971, copyright by the RAND Corporation.

recycling. Products need to be redesigned to facilitate easy recycling.

4. Build higher quality products. Increased product life would result in lower replacement rate. Initial costs of products would go up but energy would be saved.

5. Avoid producing or using throwaway products where reuse is possible. Throwaway glass bottles require an average of 4 times as much energy per unit used than do returnables. They are also twice as expensive to the consumer.

6. Use low-energy alternatives when possible. Aluminum beverage cans require 6.3 times as much energy to produce as do plated steel cans.

7. Produce chemicals from sources other than petroleum. A process has been suggested for fermenting plant carbohydrate material. The products are alcohols which can be converted to chemicals and products currently being obtained from petroleum. A great deal of development is needed to make this process feasible.

8. Produce petroleum and other useful materials from wastes. Experiments have proven that it is possible to convert animal manure into fuel oil. It has also been shown that fuel gas and oil can be obtained as products when old tires are decomposed by heating them strongly (pyrolysis).

9. Stop using natural gas to generate electricity. This dwindling supply of fuel should be reserved for more efficient use in direct

home heating. Substitute the more plentiful coal for electricity generation.
10. Avoid unnecessary industrial growth.
11. Avoid the "more, bigger, consume-at-all-costs" type of advertising.

    The facts contained in this chapter point out the great potential that exists for the conservation of energy resources. It should be recognized, however, that regardless of laws, rules, etc., conservation will take place only when an aware population undertakes the task as individuals. We must each become convinced of the necessity for energy conservation and perform the individual acts that will make it work.

Suggested Readings

Berg, C. A., "Conservation in Industry," *Science*, April 19, 1974, pp. 264–270.

Citizens' Advisory Committee on Environmental Quality, "Citizen Action Guide to Energy Conservation," Government Printing Office, Washington, D.C., 1973.

"Conservation and Efficient Use of Energy," Joint Hearings before the U.S. House of Representatives Subcommittees on Government Operations and Science and Astronautics, July 1973, Government Printing Office, Washington, D.C.

"Energy Conservation," Hearings before the U.S. Senate Committee on Interior and Insular Affairs, March 1973, Government Printing Office, Washington, D.C.

Executive Office of the President — Office of Emergency Preparedness, "The Potential for Energy Conservation — A Staff Study," Government Printing Office, Washington, D.C., Oct. 1973.

Hammond, A. L., "Conservation of Energy: The Potential for More Efficient Use," *Science*, Dec. 8, 1972, pp. 1079–1081.

————, "Energy Needs: Projected Demands and How to Reduce Them," *Science*, Dec. 15, 1972, pp. 1186–1188.

Index

1 2 3 4 5 6 7 8 9 10 –KP– 80 79 78 77 76 75 74

F